Ecological Studies
Landscape Heterogeneity and Disturbance

Edited by

W.D. Billings, Durham (USA) F. Golley, Athens (USA)
O.L. Lange, Würzburg (FRG) J.S. Olson, Oak Ridge (USA)
H. Remmert, Marburg (FRG)

Volume 64

Ecological Studies

Monica Goigel Turner
Editor

Landscape Heterogeneity and Disturbance

Contributors
D.J. Bogucki, F.H. Bormann, E.O. Box, S.P. Bratton, R. Dolan,
C.P. Dunn, R.T.T. Forman, G.K. Gruendling, G.R. Guntenspergen,
T.D. Hayes, D.H. Knight, L.A. Leitner, V. Meentemeyer,
D. Morrison, J.I. Nassauer, W.E. Odum, W.L. Pace, III,
M.M. Remillard, D.H. Riskind, P.G. Risser, D.M. Sharpe,
T.J. Smith, III, F. Stearns, M.G. Turner, R. Westmacott

With 56 Illustrations

Springer-Verlag
New York Berlin Heidelberg
London Paris Tokyo

Monica Goigel Turner
Institute of Ecology
University of Georgia
Athens, Georgia 30602
USA

Library of Congress Cataloging-in-Publication Data
Landscape heterogeneity and disturbance.
/(Ecological studies;´64)
  Includes index.
  1. Landscape protection. 2. Ecology. I. Turner,
Monica Goigel. II. Title.
QH75.L29 1987   333.73   87-4418

Typeset by David Seham Associates, Metuchen, New Jersey.
Printed and bound by Quinn-Woodbine Inc., Woodbine, New Jersey.
Printed in the United States of America.

9 8 7 6 5 4 3 2 1

ISBN 0-387-96497-5 Springer-Verlag New York Berlin Heidelberg
ISBN 3-540-96497-5 Springer-Verlag Berlin Heidelberg New York

Preface

The study of landscapes has emerged as a new discipline in ecology, focusing explicitly on spatial pattern. Ecosystem ecology traditionally focused on particular points in space and studied temporal dynamics of system components. Landscape ecology overlays on this another level of complexity, namely, the arrangement of systems in space. It examines the development and dynamics of spatial heterogeneity and its influence on biotic and abiotic processes.

Landscape pattern is generated by a variety of processes, including disturbances. In turn, the heterogeneity of the landscape may enhance or retard the spread of disturbance. Although disturbance has been an important topic in ecology during the past decade, little study has been done of the relationship between landscape pattern and disturbance. Understanding this complexity is an important task for landscape ecology, and it is the subject of this book.

This volume emerged from a landscape ecology symposium, "The Role of Landscape Heterogeneity in the Spread of Disturbance." The book presents an illustrative analysis of this topic, but does not presume to be comprehensive. The authors are drawn from a variety of disciplines, reflecting the multidisciplinary nature of theoretical and applied landscape ecology. In scope, the book includes conceptual considerations, empirical studies, and management examples. Disturbances range from air pollution to exotic species, from fire to human attitudes. The landscapes discussed are both natural and managed.

Four main sections comprise the book. The introductory chapters review landscape ecology and examine the importance of scale in landscape studies.

The second section includes six chapters that focus on specific examples of disturbances in various landscapes. The third section deals with landscape management, emphasizing landscape restoration, reserve management, and human attitudes. In the final section, conclusions are drawn and ethical implications are considered. Landscape ecology has much to gain from an infusion of ideas and perspectives from the disciplines represented in this book, along with others. I hope the reader will gain new insights about landscapes from the chapters representing disciplines other than his or her own.

Many questions remain to be answered regarding the relationship between landscape heterogeneity and disturbance. What factors control whether disturbance is enhanced or retarded as it spreads across a landscape? How should we measure resistance of landscapes to disturbances? Can we measure the relative importances of geometry and other landscape characteristics in influencing disturbance? Clever empirical studies will be required to answer these and other questions. Like looking at the world through a keyhole, we look through a small spatial and temporal window and try to understand large-scale dynamics. It is my hope that this volume stimulates questioning, testing, and expansion of the ideas contained within it, so that we can further our collective understanding of the complex dynamics of landscapes.

Acknowledgments

Chapters in this volume were externally reviewed, and I thank the reviewers for their useful insights and suggestions. I am most grateful to the following people for assisting in the review process: S. P. Bratton, R. G. Coulson, J. F. Franklin, F. B. Golley, J. M. Hartman, J. R. Karr, D. A. Kovacic, R. Lowrance, V. Meentemeyer, B. T. Milne, E. P. Odum, W. E. Odum, P. G. Risser, W. H. Romme, O. E. Sala, P. S. White, and R. G. Woodmansee. Finally, I especially thank Frank B. Golley for his guidance and continuing encouragement.

Monica Goigel Turner
Athens, Georgia

Contents

Contributors

BOGUCKI, D.J.

Center for Earth and Environmental
Science, State University of New York,
Plattsburgh, New York 12901 U.S.A.

BORMANN, F.H.

School of Forestry and Environmental
Studies, Yale University, New Haven,
Connecticut 06511 U.S.A.

BOX, E.O.

Department of Geography, University of
Georgia, Athens, Georgia 30602 U.S.A.

BRATTON, S.P.

National Park Service, Cooperative Park
Studies Unit, Institute of Ecology,
University of Georgia, Athens, Georgia
30602 U.S.A.

DOLAN, R.

Department of Environmental Sciences,
University of Virginia, Charlottesville,
Virginia 22903 U.S.A.

DUNN, C.P.

Department of Biological Sciences,
University of Wisconsin-Milwaukee,
Milwaukee, Wisconsin 53201 U.S.A.

FORMAN, R.T.T.   Graduate School of Design, Harvard
                                    University, Cambridge, Massachusetts
                                    02138 U.S.A.

GRUENDLING, G.K.   Center for Earth and Environmental
                                    Science, State University of New York,
                                    Plattsburgh, New York 12901 U.S.A.

GUNTENSPERGEN, G.R.   Department of Biological Sciences,
                                    University of Wisconsin-Milwaukee,
                                    Milwaukee, Wisconsin 53201 U.S.A.

HAYES, T.D.   Resource Management Section, Texas
                                    Parks and Wildlife Department, Austin,
                                    Texas 78744 U.S.A.

KNIGHT, D.H.   Department of Botany, University of
                                    Wyoming, Laramie, Wyoming 82071
                                    U.S.A.

LEITNER, L.A.   Department of Biological Sciences,
                                    University of Wisconsin-Milwaukee,
                                    Milwaukee, Wisconsin 53201 U.S.A.

MEENTEMEYER, V.   Department of Geography, University of
                                    Georgia, Athens, Georgia 30602 U.S.A.

MORRISON, D.   School of Environmental Design,
                                    University of Georgia, Athens, Georgia
                                    30602 U.S.A.

NASSAUER, J.I.   Department of Horticultural Science and
                                    Landscape Architecture, University of
                                    Minnesota, St. Paul, Minnesota 55108
                                    U.S.A.

ODUM, W.E.   Department of Environmental Sciences,
                                    University of Virginia, Charlottesville,
                                    Virginia 22903 U.S.A.

PACE, W.L., III   Resource Management Section, Texas
                                    Parks and Wildlife Department, % Soil
                                    Conservation Service, Plant Materials
                                    Center, Knox City, Texas 79529 U.S.A.

REMILLARD, M.M. Institute of Ecology, University of Georgia, Athens, Georgia 30602 U.S.A.

RISKIND, D.H. Resource Management Section, Texas Parks and Wildlife Department, Austin, Texas 78744 U.S.A.

RISSER, P.G. University of New Mexico, Albuquerque, New Mexico 87131 U.S.A.

SHARPE, D.M. Department of Geography, Southern Illinois University at Carbondale, Carbondale, Illinois 62901 U.S.A.

SMITH, T.J., III Department of Environmental Sciences, University of Virginia, Charlottesville, Virginia 22903 U.S.A.

STEARNS, F. Department of Biological Sciences, University of Wisconsin-Milwaukee, Milwaukee, Wisconsin 53201 U.S.A.

TURNER, M.G. Institute of Ecology, University of Georgia, Athens, Georgia 30602 U.S.A.

WESTMACOTT, R. School of Environmental Design, University of Georgia, Athens, Georgia 30602 U.S.A.

1. Introduction

1. Landscape Ecology: State of the Art

Paul G. Risser

Even casual observation reveals that most landscapes are composed of various components. A typical rural landscape might include several agricultural croplands, pastures, woodlands, streams, farmsteads, and roads. Thus, the landscape is heterogeneous, that is, consists of dissimilar or diverse components or elements. In addition to the rather obvious spatial heterogeneity, the landscape is temporally heterogeneous. Ecological processes operate at different time scales. For example, forest trees have life spans of decades, annual crops grow for less than a year, and individual stream insects may last only a few days. It is this mixture of processes consisting of different spatial and temporal scales, all operating as a system, that leads to the ideas of landscape ecology.

One has only to consider the ideas of Thoreau and the descriptions in the *Sand County Almanac* (Leopold 1949) to realize that the landscape perspective is not new. Today, practitioners of many diverse disciplines recognize the fundamental importance of landscape heterogeneity (e.g., ecology, wildlife biology, cultural anthropology, economic geography, and landscape planning). The challenge is to integrate the ideas of these disciplines into a coherent body of knowledge so that the behavior of landscapes can be understood. Thus, in a formal sense landscape ecology is the synthetic intersection of many related disciplines that focus on the spatial and temporal pattern of the landscape. As a field of scientific inquiry, it considers the development and maintenance of spatial heterogeneity, interactions and exchanges across heterogeneous landscapes, the influence of heterogeneity on biotic and abiotic processes, and the management of that heterogeneity. Concepts of landscape ecology have been developing around the world for several years (Forman and Godron 1986), but only recently has there been a formal attempt to organize these ideas within the United States (Risser et al. 1984).

Landscape ecology considers managed, as well as natural, ecosystems, and many fundamental questions in ecology and resource management require understanding the ecology of a landscape. For example, how does landscape heterogeneity affect the spread of natural and human-mediated disturbance? Aspects of this question will be addressed in the remaining chapters of this book. The objectives of this chapter are to: (1) review terms and concepts in landscape ecology, with particular emphasis on those related to landscape heterogeneity and disturbance; and (2) present a series of challenges for landscape ecologists to address.

1.1 Landscape Ecology: Definitions and Principles

In a synthetic field, such as landscape ecology, it is especially important to establish a common vocabulary. For the purposes of the discussions in this book, definitions need to be provided for the following concepts: heterogeneity, landscape, and disturbance. Heterogeneity is the simplest of the three because the dictionary definition is adequate (*Webster's Ninth New Collegiate Dictionary*, 1984): "heterogeneous: consisting of dissimilar or diverse ingredients or constituents, mixed." Of importance here is the recognition that heterogeneity applies to both spatial and temporal attributes of the landscape.

The definition of "landscape" is more problematic because landscapes can be observed from many points of view and because various scientific disciplines focus on processes operating in landscapes at different spatial and temporal scales. Forman and Godron (1986) discuss the definition of a landscape by first noting five characteristics that are typically repeated across land areas: (1) a cluster of ecosystem types, (2) the flows or interactions among the ecosystems of such a cluster, (3) the geomorphology and climate, (4) the set of disturbance regimes, and (5) the relative abundance of ecosystems within a cluster. From this observation, they define a landscape as "a heterogeneous land area composed of a cluster of interacting components that is repeated in a similar format throughout." Landscapes in this context are usually considered as items of interest at the spatial scale of tens to hundreds of kilometers, although, as subsequent discussions will show, explanations of landscape characteristics involve processes on other spatial scales.

The challenge of defining "disturbance" has plagued virtually all those who have attempted to address the topic (Barrett and Rosenberg 1981; Mooney and Godron 1983; Sousa 1984; Pickett and White 1985; Forman and Godron 1986). Forman and Godron (1986) define disturbance (or perturbation) as an event that causes a significant change from the normal pattern in an ecological system. Sousa (1984) adopts a more specific definition: "a discrete, punctuated killing, displacement, or damaging of one or more individuals" reducing the ability of the individual to survive or to become established. In a discussion that partially separates "perturbation" from "disturbance," Pickett and White (1985) define disturbance as "any relatively discrete event in time that disrupts ecosystem, community, or population structure and changes resources, substrate availability, or the physical environment." Because of its generality, the latter definition will be used in the following discussions. It is important to recognize that disturbance is a part of natural ecosystems (Pickett 1980) and does not depend on a presumption of stability (Levins and Lewontin 1985), that a portion of the heterogeneity in the landscape at any one time is caused by disturbance (Bormann and Likens 1981), and that management involves contending with and using disturbance.

1.1.1 Definition: Structure, Function, and Change

The primary focus of landscape ecology is on: (1) spatially heterogeneous areas, (2) fluxes and redistribution of materials and energy among landscape elements, and (3) human actions as responses to, and their reciprocal influence on, ecologic processes (Risser et al. 1984). To organize ideas about these focal processes, it is useful to define the fundamental characteristics of landscapes—structure, function, and change. Forman and Godron (1986) provide the following definitions:

1. Structure: the spatial relationships among distinctive ecosystems; the distribution of energy, materials, and species in relation to the sizes, shapes, numbers, kinds, and configurations of ecosystems.
2. Function: interactions among the spatial elements; the flow of energy, materials, and species among the component ecosystems.
3. Change: alteration in the structure and function of the ecological mosaic over time.

To codify the concepts for landscape ecology, it is necessary to provide a descriptive morphology of the patches of ecosystems, or building blocks, of a landscape, that is, its structure. Forman and Godron (1981) describe patches according to size, shape, biotic type, number, and configuration (Fig. 1.1). With this template, it is possible to describe patches according to a common terminology and to compare landscapes accordingly. Further, it is now possible to describe the origins of patches (Fig. 1.2) and the effects of disturbances on the configurations of these patches. This set of definitional figures provides a common set of terms for describing the elementary structural characteristics of landscapes.

Functional attributes of landscapes are not so amenable to simple descriptive terminology. However, various characteristics of the landscape significantly influence the transport and redistribution of materials. For example, studies in Maryland considered the effects of landscape pattern of agricultural cropland and riparian forest on the distribution of water, nutrients, and primary production (Peterjohn and Correll 1984). Within the nutrient budget of the landscape (Fig. 1.3), the cropland released most of its annual input of nitrogen and much

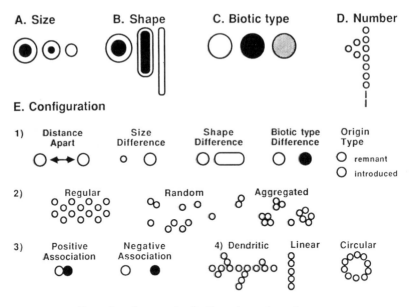

Patch characteristics in a landscape

Figure 1.1. Size and configuration characteristics of landscape patches (Forman and Godron 1981).

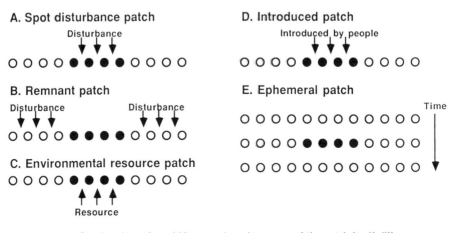

Patch origins. Species dynamics within a patch and turnover of the patch itself differ substantially according to the mechanisms causing a patch.
O O O = matrix; ● ● ● = patch; disturbance = a sudden severe environmental change.

Figure 1.2. Mechanisms that cause patches in the landscape mosaic (Forman and Godron 1981).

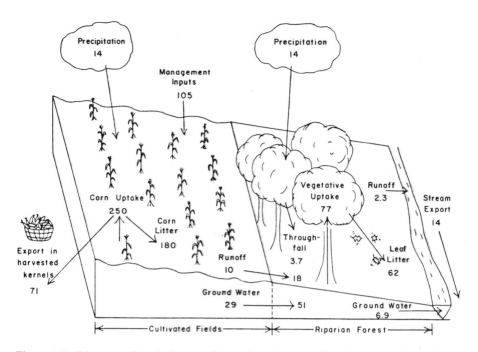

Figure 1.3. Diagram of total nitrogen flux and cycling in a Maryland experimental watershed, March 1981 to March 1982. All values are in kilograms per hectare of the respective habitats. From W. T. Peterjohn and D. L. Correll, copyright © 1984 by the Ecological Society of America. Reprinted with permission.

of the phosphorus. Conversely, the riparian forest released only approximately 10% of its annual input of nitrogen and 20% of the phosphorous. Pathways of loss of nutrients from the agricultural field to the discharge stream were primarily by groundwater for nitrogen and surface runoff for phosphorous. Of particular importance were the retention capabilities of the riparian forest. If this particular landscape were without the riparian forest, there would have been twice as much nitrate nitrogen lost to the stream. Thus, landscape patterns or patch mosaics of vegetation types can have a major influence on the nutrient dynamics of the landscape.

An important concept is that landscapes change over time but not all landscape processes occur simultaneously or at the same rate. There are three mechanisms currently posed as causal processes for development of landscape heterogeneity (Forman and Godron 1986): (1) specific geomorphologic processes occurring over long time periods, (2) colonization patterns of organisms occurring over short and long time scales, and (3) local disturbances of individual ecosystems occurring over a short time. Geomorphic processes occurring over long periods of time may influence edaphic conditions, which in turn affect both plants and animals (Schimel et al. 1985). Colonization patterns of organisms affect the biotic components of a landscape, and these components change rapidly or slowly, depending on the disturbance regime and the life history characteristics of the organisms (Pickett 1980; Armesto 1985). Examples of these myriad time scales and the concurrent changes in landscape characteristics are summarized by Delcourt et al. (1983).

1.1.2 Disturbance and Landscape Heterogeneity

The types, effects, and spatial and temporal scales of disturbances are exceedingly complex, and although the role of disturbance in ecosystems has received recent attention, little research has been conducted on disturbance at the landscape level. As summarized by Risser et al. (1984), homogeneity is often thought to enhance the spread of disturbance. Examples include the spread of pests in agroecosystems, wildfire perpetuation, spread of Dutch elm disease, and erosion. Heterogeneity, however, may also enhance the spread of disturbance, such as when small woodlots harbor white-tailed deer populations that disturb surrounding crops. Further, the effects of disturbance may increase the heterogeneity of the environment and alter the impact of a later disturbance of the same magnitude. Heterogeneity, thus, may act both as a stabilizing factor (e.g., by spreading the risk) and in fostering disturbance.

Important objectives of landscape ecology include determination of the interaction effects of heterogeneity and disturbance and the proper management of the interaction effect. The remaining chapters of this book address various aspects of these issues, and the following section of this chapter discusses several recently proposed principles of landscape ecology that have particular relevance to these topics.

1.1.3 Principles of Landscape Ecology

The Landscape Ecology Workshop held in 1983 (Risser et al. 1984) identified a number of "principles" that unite ecological and landscape perspectives, including the following:

1. The relationship between spatial pattern and ecological processes is not restricted to a single or particular spatial or temporal scale.
2. Understanding of landscape ecology issues at one spatial or temporal scale may profit from experiments and observations on the effects of pattern at both finer and broader scales.
3. Ecological processes vary in their effects or importance at different spatial and temporal scales. Thus, biogeographic processes may be relatively unimportant in determining local patterns but may have major effects on regional patterns. For example, processes leading to population decline may produce extinction at a local scale, but may be manifest only as spatial redistributions or alterations in age structure at broader geographic levels.
4. Different species and groups of organisms (e.g., plants, herbivores, predators, parasites) operate at different spatial scales such that investigations undertaken at a given scale may treat such components with unequal resolution. Each species views the landscape differently, and what appears as homo-

geneous patch to one species may comprise a very heterogeneous patchy environment to another.

5. Scales of landscape elements are defined, using spatial perspectives of sizes determined by the specific objectives of the investigation or the pertinent management question. If a study or management issue focuses at a specified scale, processes and patterns occurring at much finer scales are not always perceived because of filtering or averaging effects, whereas those occurring at much broader scales may be overlooked simply because the focus is within a smaller landscape unit.

Forman and Godron (1986) offer three principles of landscape ecology. Their Landscape Stability Principle specifies that the stability of the landscape mosaic may increase in three distinct ways, that is, toward (1) physical system stability (in the absence of biomass), (2) rapid recovery from disturbance (low biomass present), or (3) high resistance to disturbance (usually high biomass present).

In the first instance, when virtually no biomass is present in the landscape element, the system stability depends upon only the physical attributes and there are no biological means to either resist change or to assist in recovery from disturbance. When low amounts of biomass are present, there is relatively little resistance to change, but there may be a capability for rapid recovery from disturbance. Landscape elements with high biomass may resist change, but recover relatively slowly from disturbance. The Landscape Change Principle states that in an undisturbed condition, horizontal landscape structure tends progressively toward homogeneity, moderate disturbance rapidly increases heterogeneity, and severe disturbance rapidly increases or decreases heterogeneity. No landscape naturally achieves homogeneity because of the inherent differences among landscape elements and because small to moderate disturbances cause heterogeneity. Severe disturbances, of course, may either reduce heterogeneity by removing existing structure, or cause more heterogeneity by causing changes in only some parts of the structure.

Lastly, the authors suggest in their Nutrient Redistribution Principle that the rate of redistribution of mineral nutrients among landscape elements increases with the intensity of disturbance in those landscape elements. This principle is based on the concept that disturbances disrupt nutrient regulatory and conservation mechanisms, and, as a consequence, nutrients are more likely to be transferred among landscape elements or from the landscape itself.

These definitions and concepts represent much of the theoretical development of landscape ecology as it stands in the United States. Elsewhere, especially in Europe, landscape ecology has experienced a much longer gestation period and is now highly integrated into land use planning activities and decision-making arenas (Naveh 1982). In the United States, landscape ecology principles have been used in diverse fields such as wildlife management, recreation planning, and geography (Risser et al. 1984) but, as yet, have not become well integrated into routine management decisions. In addition, it is now clear that meaningful expansion of ecosystem analyses into the spatial heterogeneity of landscapes represents a powerful new approach to ecological research.

Table 1.1. Time Scales of Processes Found in a
Grassland Landscape

Process	Time Scale
Precipitation	Minutes to hours
Transpiration from canopy	Hours to days
Forage production	Days to months
Species composition changes	Months to years
Soil formation	Years to centuries
Landscape geomorphic processes	Centuries to millenia

1.1.4   Hierarchical Organization of Landscapes

One of the challenges of landscape ecology is to contend with the large arrays
of spatial and temporal scales of ecological processes and disturbance regimes
(Delcourt et al. 1983). For example, in considering the mosaic of grassland
landscapes, there are processes ranging from square millimeters to hundreds
of kilometers and from time scales of minutes to millenia (Table 1.1). The questions of landscape ecology force the investigator to confront these scaling issues.
Hierarchy theory may be a useful framework for ordering scale complexities
(Allen and Starr 1982; Allen et al. 1984). Although the theory requires more
discussion to appreciate its value, there are several points of initial importance.
First, hierarchy theory suggests that the investigator should focus on a particular
level of interest. The upper level then becomes the boundary constraint (Fig.
1.4) and can be used to explain the significance of the level of focus. Lower
levels, however, may explain the processes controlling the phenomena at the
level of focus. Second, there is no single fundamental hierarchy, but rather the
scales depend upon the phenomenon under consideration and on the viewpoint
of the investigator. Finally, in translating from one scale to another, the investigator must select a common phenomenon that changes only in grain or
detail and expanse or size of observation. Hierarchy theory provides a logical
construct around which to organize investigations of the structure and function
of landscapes.

Hierarchy theory may be particularly important in studies of disturbances
in the landscape, especially by recognizing that a disturbance at one scale may

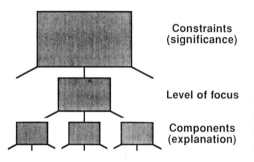

Constraints
(significance)

Level of focus

Components
(explanation)

Figure 1.4. Hierarchy diagram showing
level of forms with upper constraints
and lower levels for explanation
(O'Neill 1986).

Figure 1.5. Diagram showing how models can be used to describe and relate phenomena at different hierarchical levels.

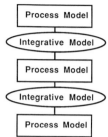

be a stabilizing force at another (Allen and Starr 1982). For example, fires adversely affect single trees, but at the scale of the fire cycle (which integrates multiple fires), fire may be essential for maintaining a mosaic of forest types or any one of the forest types (Loucks 1970). Thus, it is important to identify the pertinent scale of both the system and the disturbance.

There are three fundamental approaches to integrating the disparate spatial and temporal scales in landscapes: addition, indexes, and models. In some instances, it may be possible to simply scale from one temporal or spatial scale to another by addition. Usually this is not possible, either because the relationships are not linear or because phenomena are not defined uniformly at different scales. A second approach is to use a common index that is equally applicable at two or more scales. The third approach is to use models to characterize attributes of one landscape scale, then generate descriptors of phenomena that can be used to analyze or compare scales (Fig. 1.5). In the future, models will be used both to describe landscape phenomena and to integrate between and among spatial and temporal scales.

1.2   Challenges for Landscape Ecology

I believe we should consider the following nine challenges as we deliberate about the future of landscape ecology.

1. We must think bravely and with contemplative recklessness. We are products of the same mold of scholarship, regardless of our disciplines, and the scientific enterprise of this country tends to discourage the outlandish and the iconoclastic. Yet, advancing landscape ecology will require thinking in innovative ways that are not restricted by the extant concepts and methods of related disciplines.
2. Rejection of the troika of myths about ecological systems will lead to more realistic understanding of landscapes and ecology. These myths are: (1) that fundamental understanding about ecological systems demands that studies be conducted on natural, undisturbed systems; (2) that ecological systems fulfill the laws and expectations of equilibria without regard to scale or phenomena; and (3) that humans are intruders on ecological systems.
3. Emerging fields of science always struggle with the conflict between the

need for specific cases of observations and the inherent desire for synthetic generalizations. At this point, landscape ecology should attempt to collect both specific experimental and observational data on landscape processes while, at the same time, attempting to forge synthetic concepts.

4. These specific and general approaches to landscape ecology will profit from new analytical techniques. Although the application of existing techniques (e.g., fractal geometry or simulation models embedded in geographic information systems) will be useful, new techniques designed to address specific questions concerning landscape structure and function should be developed.

5. Reconciling the spatial and temporal scales used in studies by various disciplines to describe the same or related phenomena is the key to understanding landscape ecology. In addition, this scaling issue is paramount to the relationship between landscape and global processes.

6. The "patch" has achieved a central focus in landscape ecology, primarily because it can be visualized easily as a structural attribute of landscapes. The patch is identifiable not necessarily by ecosystem properties, but rather because it is different from the matrix or from another patch, that is, a patch is identifiable because there is a high spatial autocorrelation among the entities composing the patch. Nevertheless, patches defined in this way may have functional boundaries and, if so, a whole new set of principles can be applied: for example, membrane or boundary dynamics.

7. Many aspects of landscape ecology might be subsumed under expanded definitions of physical or regional geography. Thus, there is an imperative to examine the thoughts and ideas of geography to determine if previously established principles in geography can be usefully applied in landscape ecology.

8. There is a need to combine ecological understanding of landscapes with practical applications about land use management. European and Scandinavian countries, for example, use landscape ecological concepts to make the best management decisions for the immediate and long-term future.

9. Landscape heterogeneity is likely to be a key predictive characteristic, and disturbance represents a strong potential experimental tool to be applied under a wide range of research and management applications (Odum 1985; Risser 1985). Thus, the role of landscape heterogeneity in the spread and inhibition of disturbance is central to any possible predictive theory.

1.3 Conclusions

Further development of landscape ecology seems inevitable with the convergence of many disciplines. This diverse set of disciplines at one moment makes the field exciting, but also threatens to confound interdisciplinary communication. Also, as scientific fields mature, more commonly accepted generalities or principles begin to form the backbone of the science. Until these general principles become established, it is difficult to evaluate individual or case-by-case observations. Yet, these principles are not possible without the benefit of observations that can be eventually deductively or inductively related to the science as a whole.

To foster the development of landscape ecology, I recommend that a Portfolio of Landscape Ecology Scenarios be developed. This would be a large loose-leaf book with pages covered with plastic. The loose-leaf structure would permit convenient page substitution as new experiences are gained; the plastic covered pages would encourage use of the portfolio in the field. On each page there would be a simple diagram:

The arrow between the two compartments would be described in any number of ways, for example, simulation model, diagram, photographs, recipe for specific management actions, dichotomous key, or text. A user could file through the book searching for a description of the current landscape conditions of interest. Then there would be one or more pages describing how to understand or manage the landscape to reach the objective conditions.

Obviously, this simple portfolio would be a primitive synthesis, because most of the synthesis would be contained in the table of contents or the index. On the other hand, an enormous amount of information would be explicitly referenced, and by comparing procedures and ideas, generalities could be identified and ultimately tested in a rigorous fashion.

The above, albeit semiserious, idea of a portfolio typifies the current status of landscape ecology. Specifically, there are several case studies and a number of existing and proposed analytical and experimental techniques, all thought to be useful but not organized in a strong conceptual or theoretical manner. The present volume, however, is designed to present a number of specific examples of investigations about the landscape and disturbance. From this treatment and subsequent efforts, synthesis of landscape ideas will develop.

References

Allen, T.F.H., Starr, T.B. 1982. *Hierarchy: Perspectives for Ecological Complexity.* University of Chicago Press, Chicago.

Allen, T.F.H., O'Neill, R.V., Hoekstra, T.W. 1984. Interlevel relations in ecological research and management: Some working principles from hierarchy theory. U.S. Department of Agriculture, Forest Service General Technical Report RM-1100. Rocky Mountain Forest and Range Experiment Station, Fort Collins, CO.

Armesto, J.J. 1985. Experiments on disturbance in old-field plant communities: Impact on species richness and abundance. *Ecology* 66:230–240.

Barrett, G.W., Rosenberg, R. (eds.) 1981. *Stress Effects on Natural Ecosystems.* Wiley & Sons, New York.

Bormann, F.H., Likens, G.E. 1981. *Pattern and Process in a Forested Ecosystem.* Springer-Verlag, New York.

Delcourt, H.R., Delcourt, P.A., Webb, III, T. 1983. Dynamic plant ecology: The spectrum of vegetation change in space and time. *Quat. Sci. Rev.* 1:153–175.

Forman, R.T.T., Godron, M. 1981. Patches and structural components for a landscape ecology. *BioSci.* 31:733–740.

Forman, R.T.T., Godron, M. 1986. *Landscape Ecology.* Wiley & Sons, New York.

Leopold, A. 1949. *Sand County Almanac.* Oxford University Press, New York.

Levins, R., Lewontin, R. 1985. *The Dialectical Biologist.* Harvard University Press, Cambridge, MA.

Loucks, O.L. 1970. Evolution of diversity, efficiency and community stability. *Am. Zoologist* 10:17–25.

Mooney, H.A., Godron, M. (eds.) 1983. *Disturbance and Ecosystems.* Springer-Verlag, New York.

Naveh, Z. 1982. Landscape ecology as an emerging branch of human ecosystem science. *Adv. Ecol. Res.* 12:189–247.

Odum, E.P. 1985. Trends expected in stressed ecosystems. *BioSci.* 35:419–422.

O'Neill, R.V. 1987. Hierarchy theory and global change. *In* T. Rosswall, R.G. Woodmansee and P.G. Risser (eds.), *Spatial and Temporal Variability in Biospheric and Geospheric Processes. SCOPE Report.* Wiley & Sons, Chichester, England.

Peterjohn, W.T., Correll, D.L. 1984. Nutrient dynamics in an agricultural watershed: Observations on the role of a riparian forest. *Ecology* 65:1466–1475.

Pickett, S.E.T. 1980. Non-equilibrium coexistence in plants. *Bull Torrey Botan Club* 1007:238–248.

Pickett, S.T.A., White, P.S. 1985. *The Ecology of Natural Disturbance and Patch Dynamics.* Academic Press, New York.

Risser, P.G. 1985. Toward a holistic management perspective. *BioSci.* 35:414–418.

Risser, P.G., Karr, J.R., Forman, R.T.T. 1984. *Landscape Ecology: Directions and Approaches.* Illinois Natural History Survey Special Publication, No. 2.

Schimel, D., Stillwell, M.A., Woodmansee, R.G. 1985. Biogeochemistry of C, N, and P in a soil catena in the shortgrass steppe. *Ecology* 66:276–282.

Sousa, W.P. 1984. The role of disturbance in natural communities. *Ann. Rev. Ecol. System.* 15:535–591.

2. Scale Effects in Landscape Studies

Vernon Meentemeyer and Elgene O. Box

2.1 Introduction

Landscape ecology cannot escape dealing with spatial analysis, spatial scale and scale-change effects. A landscape may appear to be heterogeneous at one scale but quite homogeneous at another scale, making spatial scale inherent in definitions of landscape heterogeneity and diversity. In analyzing disturbances and other aspects of landscape change, temporal scale (or temporal resolution of events) may also become an important factor, for similar reasons. The importance of integrating the disparate spatial and temporal scales in landscapes was emphasized by Risser in the preceding chapter.

Ecologists now also seem to be thinking more and more about different spatial and temporal scales. This is suggested, for example, by such relatively new phenomena as the programs for Long-Term Ecological Research (LTER, e.g., Michener 1986), discussion of a new International Geosphere-Biosphere Program (IGBP) to aid integrative research at global scale (e.g., NASA 1986), and increased interest in what has come to be called landscape ecology (e.g., Tjallingii and de Veer 1981; Risser et al. 1983; Vink 1983; Naveh and Lieberman 1984; Forman and Godron 1986; Schreiber 1986). The IGBP discussions, in particular, have recognized a need for studying problems associated with the various spatial and temporal scales at which data are gathered (St. Petersburg mtg. 1985). Landscape ecology, first identified by Troll (1939, 1950, 1966, 1968), arose out of European traditions of landscape-level geography and vegetation science (e.g., Tüxen 1968), which recognized scale problems somewhat earlier.

In North America scale problems (outside cartography) have perhaps been most familiar in planning activities (e.g., McHarg 1971; Clark 1976), in the statistical concept of spatial autocorrelation (e.g., Cliff and Ord 1973), in soil science and geomorphology (e.g., Campbell 1977; Imeson 1984), in systems theory (e.g., Margalef 1968; Allen and Starr 1982; Salthe 1985), and in more traditional vegetation and population-level field studies (e.g., Kershaw 1970; Hill 1973; Pielou 1975; Wiens 1976; Watts 1984). More recently, scale has become recognized as an important factor in ecological studies focusing on disturbance (e.g., White 1979; Mooney and Godron 1983; Pickett and White 1985), forest dynamics including implications for habitat preservation and species conservation, (e.g., Johnson and Sharpe 1976; Burgess and Sharpe 1981; Sharpe et al. 1981; Harris 1984), body size (Peters 1983), and urban geography (e.g., Dorney and McClellan 1984).

Temporal scale has been considered in studies of climate change (e.g., Hansen et al. 1985; Lewin 1985), and of evolution and historical biogeography (e.g., Flenley 1984). Delcourt et al. (1983) have related spatial and temporal scales as they apply to vegetation dynamics. DeAngelis et al. (1985) show how scale can affect the properties of a system, including its degree of equilibrium, degree of closedness versus openness, and whether a disturbance agent is endogenous or exogeneous. Even human ecology (s.l.), in the course of describing the "subversive" nature of ecology (e.g., Shepard and McKinley 1969; Ophuls 1977), has not overlooked some indirect implications of scale, including such concepts as "economies of scale" and laws of "diminishing returns." Despite this and other recent interest in scale problems (e.g., Miller 1978; Romme and Knight 1982; Bailey 1985), comprehensive theories of scale have not really emerged.

The purposes of this chapter are (1) to review scale as it relates to the study of landscape diversity and disturbance, as well as to the whole field of landscape ecology; (2) to emphasize the role of scale in research designs; (3) to examine what may be needed to develop a science of scale; and (4) to suggest scale-related principles and hypotheses that may foster and/or facilitate ecologic study at the landscape level. Instead of a more analytical treatment of fewer issues, we have tried to collect as many as possible of the miscellaneous "ingredients" which may form a conceptual framework for more specific studies and for more general theories of scale and scale-change effects. This approach is quite incomplete and does not possess the maturity, for example, of Forman and Godron's (1986, pp. 24–28) seven general principles of landscape ecology. But we may already know much more about landscape ecology than we do about scaling and scale-change effects in general. As a result, this broad but rather unsatisfying initial cataloguing may be necessary, especially when phenomena are very complex and interrelated, some basic concepts are still being sought, and most other work is more highly focused on particular situations.

We begin by looking at various scale aspects from geography, the spatial science, which has a long history of landscape studies. The emphasis is on spatial scale, but temporal scale is also considered. Scale study involves many disciplines, and this review represents our experience in geography and ecology, including experience with broader (coarser) scales and with abiotic factors as much as biotic.

2.2   What Is Scale?

In cartography, spatial scale is a well-defined concept representing the degree of spatial reduction, usually expressed in terms such as 1:10,000. The scale on a map, however, may not be the same everywhere. Commonly it changes with latitude, depending on the type of projection used to turn the three-dimensional earth surface into a two-dimensional map. For small areas this is not a problem, but all maps of large areas have distortions in distance, direction, area, and/or shapes, depending on the projection used.

The purpose of a map is to convey information, and the level of detail to be included on a map is often a more difficult decision than the selection of the correct projection. Excessive information may create useless clutter, whereas a map with too little detail conveys little information. Similarly, in a landscape small and/or homogeneous regions may contain too little "information" to be interesting, whereas large and/or complex regions (or landscapes) may contain more information than can reasonably be processed (e.g., Bailey 1985).

Most ecological studies in which the spatial dimension is important have used *absolute* scale, which involves distance, direction, shape, and geometry. *Relative* scale, on the other hand, transforms space based on some functional relationship. For example, a map of driving time between places might show places as far apart if driving time is long (worse roads, heavy traffic, etc.) and close together if driving time is short. Brown and Gersmehl (1985) use such a space-time transformation to represent the dispersal of grasses in North America. Another useful basis for space transformation might be expenditure of energy or time plus energy (see Fig. 2.1; Muehrcke 1980). Places would be relatively "close," then, if travel between them requires little energy expenditure. Relative scale might be useful in studying the propagation of disturbances across landscapes.

Temporal scale is much like spatial scale except that time involves only one dimension and one direction. An important aspect of temporal scale in ecology is that many processes, including important changes, proceed slowly and require long periods of observation (hence, the recent establishment of LTER networks in the United States). Processes such as growth of herbaceous vegetation, herbivory, and loss of liquid photosynthates, on the other hand, may proceed so rapidly that measurement accuracy decreases rapidly with coarser temporal resolution (e.g., Dickerman et al. 1986).

Special problems are posed by needs to interface results and data gathered at different temporal or spatial scales. This very common problem requires adequate planning and some acceptance of standard methods.

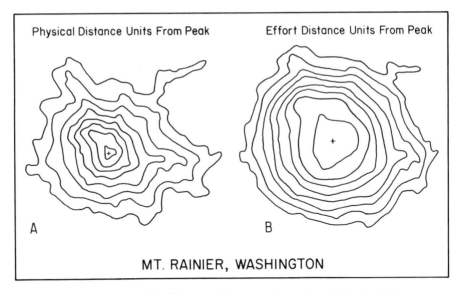

Figure 2.1. "Relative scale," as illustrated by a transformation of physical distance into a climbing effort distance for Mount Ranier, Washington (Muehrcke 1980).

2.3   Effects of Scale Change

It is important to understand and anticipate the implications of moving from one scale to another. Perhaps the most obvious effect of scale change is the level of discernible or treatable detail that is present. Moving to a coarser scale means that one is further removed from the basic processes. The detail is still present but not in a form that conveys relevant information. Table 2.1 presents a list of some common effects of scale change. One method of compensating

Table 2.1. Some Effects of Changing Spatial Scale on Analyses and Model Structures

Phenomenon	Size of Study Area	
	Small	Large
Level of discernible or "treatable" detail	Higher	Lower
Potential for experimental manipulation	Higher	Lower or none
Number of important factors or variables	More	Fewer
The "important" variables	Chemical/ biotic	Abiotic (especially physical)
Value ranges of the landscape variables	Smaller	Greater (but depends on the variable)
Importance of the temporal scale	More	Less
Emergent properties	Fewer	More

for the problem of scale change is to isolate a few variables that vary spatially and concentrate on their geography.

Finer scales are the goal of reductionist science, in part because of the dominance of the laboratory in the late 20th Century. Up to a point, smaller areas generally permit more experimental manipulation of the thing/process under investigation, whereas larger areas are more difficult to manipulate. Furthermore, there is a space-time convergence such that observation of many processes in small areas or at very fine scale can often be made in a short time in the laboratory. Coarse scales and larger areas often involve long observation periods because large-area systems appear to have long relaxation times.

The number of variables useful for modeling and other analyses also changes with scale, generally becoming smaller at coarser scales. For example, consider the prediction of the spatial patterns of litter decomposition rates on the basis of the chemical/physical properties of litter and the moisture and energy supplied by the macroclimate (e.g., Meentemeyer 1978, 1984; Dyer 1986). At a particular site, most of the variation in decay rates for different litter species can be explained by litter properties. As one moves to broader spatial scales, the litter properties became less important, at least in the framework of regression models, and the climatic variables begin to account for most of the variance in litter weight loss. At the global scale only climate is necessary to predict the discernible patterns of many ecological phenomena, including not only litter decomposition but also litter production (Meentemeyer et al. 1982), soil carbon (Meentemeyer et al. 1985) and carbon dioxide (Brook et al. 1983), primary productivity (Lieth 1975; Box 1978), and general vegetation structure (Box 1981). Furthermore, at broader scales, even climatic variables reduce to only a few useful types. These are mainly indices of: (1) energy or heat availability, (2) moisture availability, and (3) seasonality (e.g., Dyer 1986). At the scale of the entire United States, climate appears to explain the broad general patterns of soil pH (Folkoff et al. 1981), soil base saturation (Wojcik et al. in press), and general soil type, while masking local variation that may be enormous.

Studies of total nitrogen discharge *per hectare* for a watershed have shown that, at the level of the individual watershed and its surrounding watersheds, many variables (e.g., land use, fertilizer applications, slopes, soil types, etc.) have a control on total nitrogen discharged per unit area. At the scale of the entire temperate zones of Earth, however, there appears to be only one dominant control, precipitation and volume of runoff per unit area (Meentemeyer, unpublished). Thus, at the continental or broader scale, some reasonable estimates of nitrogen balance can be made on the basis of only measured or estimated runoff and nitrogen in precipitation.

Scale changes may also dramatically change model structures, since (as just stated) the number of "important" variables generally decreases toward broader scales. This switch occurs in the models because some variables change greatly with a change in spatial scale, whereas others do not. In the litter decomposition studies mentioned earlier, the ranges of the litter quality variables increase very little toward the broader scales, but the ranges of the abiotic, especially climatic, variables increase greatly.

Spatial scale also seems to temper the importance of time in model structures. Time is needed to determine rates, but at extremely fine scales measurements of a few hours may be sufficient to determine rates of movement, dispersal, or other action. At fine scales, one sees a minute portion of the entire system. Presumably, if one were small enough, one could see the movement of an electron in a few picoseconds. Large-area systems seem to progress slowly, so much longer periods of observation are needed. Thus, at broad scales, observations may need to be conducted for many years. At broad scales, efficient and relatively accurate models may require identification of only a few constraining variables. Rates determined from measurements over a few hours, however, probably will not be as relevant over broader scales.

Studies of the temporal and spatial variability of precipitation events clearly show space-time relationships (Landsberg 1983). Generally, the shorter the precipitation event, the greater the gauge density required for a given level of measurement accuracy (see Fig. 2.2). Precipitation gauges only 5 km apart may have correlation coefficients of less than 0.5, especially for summer storms, whereas extrapolation to distances of 10 km and more often yields negative correlation. Because space-time sampling and the rules for extrapolation and interpolation are well developed in hydrometeorology (e.g., Wiesner 1970), these rules may also be useful for landscape ecology. These rules, however, may change seasonally and geographically.

Although moving to coarser scale may result in loss of detail, it also may result in the appearance of emergent properties due to synergisms at a higher level of system integration. These interactions were not seen at finer scales

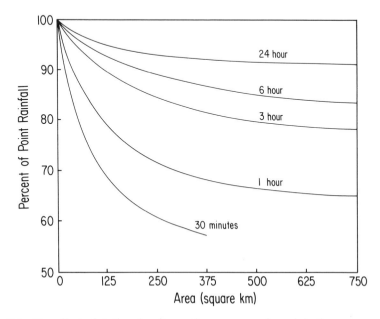

Figure 2.2. The effect of station density on the accuracy of precipitation measurements.

and were perhaps not acting if finer scale meant a lower level of integration. Hierarchy theory (e.g., Allen and Starr 1982), thus, may have an important bearing on scale, because systems are usually naturally nested in space and time. As one moves up a hierarchy (usually meaning to larger areas and/or longer periods of time, i.e., coarser scales), the overall constraining variables become more apparent and/or dominant (and fewer, as noted earlier).

Finally, spatial scale changes should not be viewed as simple linear change. Doubling of map scale, for example, results in a fourfold increase in the size of the map needed to cover a given area. Haggett et al. (1965) proposed a logarithmic system for defining scale based on factors of 10 and on the size of Earth. This scale, termed the G-scale, is similar to the pH scale or the Richter scale and can be used to define any spatial scale including that of landscapes.

2.4 Impact of Spatial Scales on Analyses

In science the raw data provide the finest level of resolution obtainable, so the analysis must be compatible with their implied scale. The scale of an analysis may be dictated, as well, by other aspects of scale in the real world. Several such constraints on the scale of analysis are suggested in Table 2.2.

Patch density and structure may represent a very important constraint. If patch density is high, a relatively smaller area may suffice for the intended analysis. Haggett et al. (1977) state, "The smaller the unit of regional study [i.e., a patch] the greater the population size from which samples may be drawn. Conversely the larger the unit, the fewer the cases there are to compare." Studies of forest gaps could probably be studied in an area of a few square kilometers if gap density is large. Geographic models are rarely based on the observation of just a few regional units and must be geographically general (i.e., function well in all types of regions) in order to be generally useful. Much of what we have learned about patch dynamics, disturbance regimes, and landscape heterogeneity, however, seems to have come from the observation of a few areas and often a few patches.

The functional linkages of properties and interactions may also dictate the scale of analysis. The scale must match the process. Occasionally, the process of interest seems to occur so slowly and over such a large area that observations

Table 2.2. Factors Influencing the Selection of Spatial Scales for Analyses

Patch density (i.e., structural and spatial heterogeneity)
Nature of functional linkages or interactions of interest
Usefulness of parametric statistical methods
Structural detail and resolution
Speed of processes involved
Need to use a place-for-time substitution
Need to manipulate a system experimentally
The chance of finding emergent properties

are not practical. Broader scales usually involve more absolute distance (although not relative distance), more time, and perhaps more energy expenditure. Therefore, it may be necessary to substitute different places for different points in time. Much research on ecological succession has used place-for-time substitutions. Knowledge of when a fire or some other disturbance occurred, or when a field was abandoned, often yields useful information. Yet guidelines for when and how to make this substitution, or for the appropriate distance between the observation sites, that is, patches, have not been formulated. Presumably, large separation distances may also involve large differences in the abiotic environments.

The study of broad areas usually precludes experimental manipulation of the system, although natural experiments, such as the Mount St. Helen volcanic eruption, may be useful. Consequently, the scope of the study is usually dictated by the availability of actual data and the degree to which they represent the geography of the study area (Box 1986). More generalized information can also be derived from aerial and satellite photography, topographic maps, and soils maps. Scale-related trade-offs often involve the size of the area to be investigated, the resolution of the data, its appropriateness for the study, and the volume of data to be manipulated. Spatial scale is a last consideration in many studies, yet scale often selects the nature of the results.

2.5   A Science of Scale?

To determine the feasibility of a science of scale, one might begin by looking at the "ingredients" of sciences in general (see Table 2.3). A science can be defined by its subject matter (phenomena), vocabulary, measurement and analytical tools, and the nature of its hypotheses and questions. Some scale-related concepts and terminology, as might be related to landscape ecology and to studies in disturbance and heterogeneity, are listed in Table 2.4, including commonly used terminology unique to the discipline.

Geographers often claim that a study becomes geographical when the spatial dimension is an explicit variable in the analysis. This might include distance, area, direction, shape, and spatial correlation, as well as relative or transformed space variables. Consequently, it seems that *a science of scale must include scale as an explicitly stated variable* in the analysis. If a phenomenon is investigated at several scales, to see the change in resulting generalizations or models, its study might benefit from considerations of general scale principles.

As examples, one might consider the following. Often a spatially nested

Table 2.3. Ingredients of Any Science

Phenomena	Questions
Concepts/vocabulary	Hypotheses
Measurement methods/tools	Principles
Analytical tools	Paradigms
	Application methods

Table 2.4. Scale-Related Concepts and Terminology Relevant
to Landscape Ecology

Scale	Absolute/relative density
Scale change	Spatial structure (i.e., diversity, heterogeneity)
Distance decay	Temporal structure
Patch	Pattern
Resolution	Functional linkages
Emergent properties	Place, site, region
Hierarchy	Feedback
Disturbance	Relaxation time
Persistence/disappearance	Mapping
Change (direction, speed, variability, etc.)	System (general, spatial, spatio-temporal)
Residence/turnover time	Ecotope

sampling procedure can be used, as was done by Edmonds et al. (1985) for soil characteristics, or a within-habitat versus between-habitat analysis, as was done by Caffey (1985) for intertidal barnacles. Scale is implicit in many studies of territory and feeding patterns across a landscape (Jones and Krammel 1985). Furthermore, scale-related issues may be apparent in large oceanic ecosystems, as shown by Steele (1977) for plankton communities, as well as smaller terrestrial systems. Seabird patchiness in the open ocean shows scale-dependent patterns (Schneider and Duffy 1985). The scale of heterogeneity was found to influence foraging behavior of bumble bees (Plowright and Galen 1985). In uniform habitats, bumble bees move much more quickly and are less likely to backtrack than in habitats with boundaries and landmarks. Thus, for bumblebees, the "relative" distance of a uniform area is less than for a heterogeneous area.

2.6 Tools for a Science of Scale

Most techniques for measuring scale phenomena involve spatial scale. A partial list of potential measures is given in Table 2.5. Some of these methods can also be used for temporal scale, especially the more purely statistical methods.

Table 2.5. Potential Measures for Including Spatial Scale in Landscape Studies

Cartographic scale (a. absolute; b. transformed or relative)
Transect and quadrat data, especially when derived from spatially nested hierarchies
Resolution measures, especially as used in aerial photointerpretation and remote sensing
Diversity measures (structure, function, variability)
Measures of connectivity and landscape structure
Levels of organization—hierarchy trees
Statistical measures of clustering and coefficients of geographic association
Measures of within- and between-patch variability
Fractal dimensions

Table 2.6. Analytical Tools for Studying Scale Effects in Landscapes

Tests for spatial autocorrelation and tests for spatial patterns in residuals
Analysis of spatial, temporal, functional, and typological diversity patterns
Textural analysis and fractal geometry
Interaction matrices, connectivity models, and graph theory
Hierarchical analysis
Geographic Information System (GIS)
Application of compartment-transfer methods to changing areas in patch, matrix,
   corridor, etc. (including change probabilities)
Information theory and entropy methods
Multivariate statistical analyses
Statistical expansion methods, or "Casetti methods," useful for making regression
   models applicable across a great range of geographic scales (Casetti 1970)
Place-for-time substitutions
Hydrometeorology methods
   Arithmetic means
   Areal weighting polygon
   Triangulation methods
   Linear isohyetal
   Subjective isohyetal
   Revised weighting
Weighted interpolation methods, including Krieging

More analytical methods for representing and studying scale-related phenomena are listed in Table 2.6. None of these seem to be unique to landscape
ecology in general or disturbance and heterogeneity in particular. Traditional
multivariate statistical methods must be used with great caution in landscape
studies. Haggett et al. (1977) stress that "location data are generally spatially
autocorrelated, non-stationary, non-normal, irregularly spaced and discontin-
uous." In other words, spatial data violate virtually all rules for parametric
statistical analyses. Usually we have little information on the level of autocorrelation in geographic variables. Yet, without spatial autocorrelation there
would be no pattern, contrasts, or boundaries in the landscape. Some of these
problems can be overcome by testing for the spatial patterns in residuals and
by going back to the original observations.

Interaction matrices and connectivity models are useful (Forman and Godron
1986) as are diversity indicies when applied to landscape rather than species
diversity (e.g., Romme and Knight 1982; Shugart and Seagle 1985; Hoover 1986).
Textural analysis is becoming more important in vegetation studies, especially
in Europe (Barkman 1979).

Soil scientists and geographers have both developed "geostatistical" methods
which take into account spatial autocorrelation. The theory of regionalized variables, developed for mineral mining and exploration (Matheron 1971), has been
applied in many fields, including hydrogeology, bathymetry, and meteorology.
McBratney et al. (1982) applied these methods to agronomic data and problems,
and they reviewed the concept and its application. Although designed to treat
spatial autocorrelation, these methods are also potentially useful for finding

and quantifying scale effects. McBratney et al. (1982), for example, found variation in copper and cobalt in Scottish topsoil at several scales, including (1) an agricultural or field-to-field variation, and (2) a more distant variation probably caused by geology. A "nugget" effect ("nugget variance") caused by variation within the smallest measurement distance of less than 200 m also was evident.

Other analysis methods which involve spatial autocorrelation include the "Krieging" methods used to improve interpolation and the drawing of isolines. These methods are used mostly by geologists and soil scientists at present. A good example is found in Dahiya et al. (1985) for the spatial variability in soil nutrients. (Also see Sharpe et al., Chap. 8, this volume.)

Spatial analysis methods based on information theory and entropy have been used by geographers to model spatial concentration and diffusion processes. An approach which is even more complex, but which shows good potential, is the use of fractal geometries and fractal dimensions to model landscapes.

Geographic Information Systems (GIS) are becoming popular because they can be used to manipulate large volumes of spatial data and formulate new hypotheses about spatial patterns. Kesner (1984) used a GIS based on the Map Analysis Package (Tomlin 1980) to study the spatial dynamics of nitrogen in an agricultural watershed in south Georgia. Furthermore, a GIS with data of proper resolution could be used to calculate spatial autocorrelation, plot residuals, and manipulate scale to assess the effects of scale changes.

Finally, it is clear that the measurement instruments themselves may be scale and hence spatially biased. Roels (1985) examines the problems and apparent deficiencies in the use of catchment troughs to estimate erosion of larger areas. Thornthwaite and Mather (1955) show that the measurement of evaporation by a very small evaporating surface, as in the piche evaporimeter, can produce exceptionally biased results (Fig. 2.3). Very small evaporating surfaces better

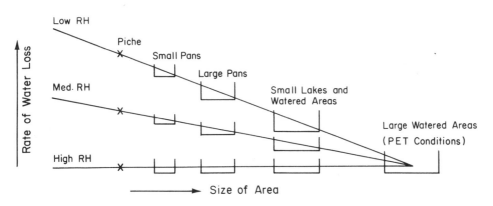

Figure 2.3. The "oasis effect" in measurement of evapotranspiration (Thornthwaite and Mather 1955).

represent the humidity of the atmosphere than the true evapotranspiration rate of a vegetated area or evaporation of a lake. It is time to examine the hardware used in field and laboratory research for this "piche" type of scale bias.

2.7   Toward Principles of Scale

The formulation of principles is dangerous when it is based on experience rather than the results of a carefully circumscribed piece of research. Furthermore, it is virtually impossible to give proper credit to original works which have become part of that experience. Nevertheless, in Table 2.7, we attempt to give an initial list of what appear to be some general principles involving scale or scale-related phenomena. Several items overlap our earlier discussion, some need expansion, and some need no comment.

The rule that everything is related to everything else, but near things are more related than distant things, has been termed Tobler's first law of geography (Tobler 1969). Distance decay is a spatial concept, similar to diffusion, but it also occurs at many scales. The distance and density of propagule dispersion from one patch to another may show a distance-decay pattern of diffusion, although heterogeneity and disturbance patterns may alter the coefficients.

Wiens et al. (1985) discussed the role of boundaries in landscape studies and recognized that boundaries may retard the flow of certain animals, propagules, and nutrients but not others. They also emphasized the powerful control by edaphic factors on boundaries and their properties. Edaphic factors should help perpetuate patches, but the geographic scale at which edaphic factors enter and leave a model or "explanation" of patch dynamics is usually unknown.

Miller (1978) provides examples of patch persistence caused by the ability of the patch to create and perpetuate its own abiotic environment. It seems

---

Table 2.7. Some Hypothesized Principles of Scale and Scale Change

An increase in size of study area tends to increase the range of values for a landscape variable

Fewer variables may be needed for modeling larger study areas; conversely, smaller study areas may have more external effects, thus requiring more variables

The smaller the study area, the greater the potential for experimental manipulation

Larger phenomena have longer relaxation times

Dynamics observed at finer scales cause the equilibria observed at broader scales

Distance decay: everything is related to everything else, but near things are more related than distant things (Tobler's first law of geography, 1969)

Greater organization yields greater potential for information flow (Margalef 1968)

Apparent detail is lost with increased area (subject to the density of patchiness); decreased area involves newly apparent detail

Increased system size involves emergent properties; decreased size loses some interactions and thus some functional properties

Patch density and appropriate scale of analysis are inversely related

Areas of dense patchiness must be analyzed at relatively finer scales than the rest of the study landscape that larger patches have a greater ability to create their own "weather." Even small patches, such as wind breaks, can increase their moisture supply by capturing snow.

2.8 Current Problems in Ecology

Some other, perhaps rather important examples of scale-related phenomena and consequent methodological problems can be seen in a few other, selected areas of current interest in ecology. An obvious instance of scale problems in ecological data gathering involves remote sensing (e.g., BioScience 1986). The spatial resolution of LANDSAT images is greater than for images from the Advanced Very-High-Resolution Radiometer (AVHRR) on NOAA satellites. Nevertheless, the AVHRR is more useful in global ecological studies because it passes over every part of the earth each day (greater temporal versus spatial resolution). With AVHRR it is possible to get enough cloud-free days each month to develop complete annual global sequences of monthly data, such as for the normalized difference vegetation index (NDVI) now widely used to estimate biomass and primary production (e.g., Tucker et al. 1985). Verification of the relationships between remotely sensed pixel greenness and actual primary production, however, requires actual measurements of primary production. Comparison of the two types of data quickly meets with problems of unrepresentative or "mixed" pixels that are too coarse to capture local spatial heterogeneity, especially at small measurement sites in mountains or near-coastal areas.

Scale problems also appear in the measurement of photosynthesis and primary production, both major interests of the International Biological Program (1964–1974) and subsequent ecosystem-level research. Photosynthesis is usually measured at different times during a day in units of grams of carbon dioxide per unit of leaf area per unit of time. Supposedly, the leaves or groups of leaves used are representative of the system. Primary production and respiration are normally measured in easily converted units of grams dry matter (or carbon or carbon dioxide) per unit ground surface area rather than leaf area, and per day or some longer time unit such as an entire year. Measured photosynthesis rates can vary drastically over short distances and time scales (e.g., from hour to hour). Consequently the spatial and temporal scales of the many photosynthesis measurements (e.g., Cooper 1975; Hesketh and Jones 1980) render them almost useless for estimating productivity. Means need to be developed to extrapolate those measurements to more general scales. With interest now in more global studies, ecologists have quickly identified that even good productivity data from site measurements are hard to generalize to even moderately larger areas or to longer time frames.

The potentially formidable nature of scale problems in data-gathering activities is also well illustrated by attempts to gather phenological data where temporal resolution is essential. In one of the very few attempts to relate foliation phenology (as opposed to flowering) to climatic data over a large region, Wick-

ham (1984) found that the variability in temperature data, and in the dates recorded by his phenological observers throughout the eastern United States, pre-
cluded all but the most general conclusions. Apparently one tree or a few well-
selected trees may not represent the phenological patterns over even a few
square kilometers.

2.9 Additional Questions

Our initial list (Table 2.7) of possible principles, even if valid, leaves some large
gaps in the knowledge base needed for dealing with scale phenomena, in landscape ecology and in general. Within the context of landscape ecology, some
additional questions are stated in Table 2.8. Most of these questions are non-
quantitative but rather deal with entitation, classification, cybernetics, or emergent properties or detail. For example, have we found any general principles
for the effects of landscape structure on the spread of disturbance (or vice
versa)? Can we develop any general principles for predicting or dealing with
emergent properties? Do patches or other landscape elements select (for con-
tinuation or demise)? If so, then this suggests the importance of developing a
spatial perspective on cybernetics, a sort of "landscape cybernetics." Perhaps
Garrett Hardin's well-studied "tragedy of the commons" (see also Hardin and
Baden 1977) provides an instructive example. And finally, what should we
measure in experimental or field studies of landscape dynamics?

The foregoing tables and discussion have not exhausted the list of apparently
scale-related phenomena that have been suggested by various studies, examples,
or speculation. Some of these seem to represent general patterns, at least within
a certain range of situations. As a result, some additional scale-related relationships are listed in Table 2.9. These represent "hypotheses" at this point
and are offered, without further support, for other researchers to corroborate
or deny. Cybernetic relationships seem important in many of these hypotheses.
Counter-examples will deny global validity, but there may remain limits within
which some of these hypotheses are valid and useful.

Table 2.8. Some Remaining Scale-Related Questions

How do mathematical and/or statistical properties of surfaces vary with scale?
What are patches, boundaries, etc.?
Do patches select for their own preservation or demise? If so, how, why, and at
   what scale?
What should be measured in studies of patch dynamics, scale change, disturbance
   effects, etc.?
How can we measure variable processes rapidly enough? How can we measure fast
   and slow processes simultaneously?
What are catastrophic events, and how are they related to scale?
What are the effects of landscape structure on spread of disturbance and vice versa?
What are the effects of (scale of) disturbance on landscape cybernetics?
How can we identify, measure, and analyze spatiocybernetic aspects of landscapes?
Are there principles for prediction of emergent properties in landscapes?

Table 2.9. Additional Scale-Related Hypotheses

Small patches have shorter lifespans; big patches create their own environments and
   persist better (Miller 1978)
Scale of heterogeneity is regulated by animals more than plants and may be related
   to the size, vagility, and gregariousness of the animals
Heterogeneity speeds processes within landscapes (between components) but may
   slow I/O processes of whole landscapes versus the outside environment (e.g.,
   runoff)
Scale of structural heterogeneity dictates the scale of the main functional linkages
Structure is usually the result of several important processes; thus studies of
   structure require finer scales than studies of the formative processes
Studies of cybernetic phenomena require larger scope than studies of individual
   components/fluxes (i.e., segments of a whole feedback system)
Greater separation in time and/or space (e.g., externality, time lag) between the
   cause of a disturbance and its perception in human systems reduces the cybernetic
   self-regulation by the causal agent
Larger accumulations of structure, energy, or matter have greater potential for
   regulating landscape-level dynamics
Disturbance may enhance homogeneity in (abiotically?) homogeneous environment
   (e.g., fire in flat versus complex terrain)
Spatial heterogeneity may enhance other types of heterogeneity
Environmental heterogeneity reduces direct (resource) competition (thus competitive
   exclusion), thereby enhancing species diversity
There is no fundamental level of investigation (O'Neill)

Perhaps the most ambitious and complete step toward a science of scale is
the book *Hierarchy* by Allen and Starr (1982). In a sense, this book begins
where Margalef (1968) and Weinberg (1975) leave off, as mentioned in the introduction. Allen and Starr do not organize lists of "principles" as boldly
(recklessly?) as we have done in Table 2.7, but their book contains a wealth
of suggestions and examples for development and testing of hypotheses in landscape studies and ecology in general. Their book culminates in the final chapter
on "Scale as an Investigative Tool," in which scale is treated as a potential
unifying concept (paradigm) for treating structure and behavior. This has implications for data collection, analysis, and modeling. We recommend Allen
and Starr as a somewhat difficult but very useful companion to other treatments
of scale problems (e.g., Miller 1978) and of landscape ecology in general.

   In the first chapter of their new book *Landscape Ecology*, Forman and Godron (1986) suggest seven "emerging general principles" of landscape ecology.
These principles seem to be a good initial basis for landscape ecologic studies.
If they are general principles, then they must hold at all relevant spatial and
temporal scales, and in all different regions and environments. Of course, re-
gionalization is also scale-related, requiring different criteria and perhaps resolution, depending on the nature of the terrain. Furthermore, although the species appears to be the appropriate biological unit when looking at a single
landscape, comparison of landscapes in different regions requires more functional units, such as life forms or vicariant species groups. This complicates
the development of truly general landscape principles. Nevertheless, subjecting

hypothesized landscape principles to scrutiny at various scales and in various environment types appears to be the most effective, systematic way of "validating" or disproving such hypotheses. Such exercises may also provide additional insight into scale effects and the validity of our own hypothesized principles of scale.

References

Abramsky, Z., Rosenzweig, M.L., Brand, S. 1985. Habitat selection of Israel desert rodents: Comparison of a traditional and a new method of analysis. *Oikos* 45(1):79–88.

Allen, T.F.H., Starr, T.B. 1982. *Hierarchy*. University of Chicago Press, Chicago.

Arsufti, T.L., Suberkropp, K. 1985. Selective feeding by stream caddis fly (Trichoptera) detritivores on leaves with fungal-colonized patches. *Oikos* 45(1):50–58.

Bailey, R.G. 1985. The factor of scale in ecosystem mapping. *Environ. Management* 9:271–276.

Barkman, J.J. 1979. The investigation of vegetation texture and structure, pp. 123–160, *In* M.J.A. Werger (ed.), *The Study of Vegetation*. Dr. W. Junk bv Publishers, The Hague.

BioScience. 1986. Ecology from space. *BioSci.* 36(7): special issue.

Box, E.O. 1978. Geographical dimensions of terrestrial net and gross primary productivity. *Radiat. Envl. Biophysics* 15:305–322.

Box, E.O. 1981. *Macroclimate and Plant Forms: An Introduction to Predictive Modeling in Phytogeography*. Tasks for Veg. Science, Vol. 1. Dr. W. Junk bv Publishers, The Hague.

Box, E.O. 1986. Data problems in large-area ecological modeling, pp. 247–262. *In* W.K. Michener (ed.), *Research Data Management in the Ecological Sciences*. University of South Carolina Press, Columbia.

Brook, G.A., Folkoff, M.E., Box, E.O. 1983. A world model of soil carbon dioxide. *Earth Surf. Processes* 8:79–88.

Brown, D.A., Gersmehl, P.J. 1985. Migration models for grasses in the American mid-continent. *Ann. Assn. Am. Geographers* 75:383–394.

Burgess, R.L., Sharpe, D.M. (eds.) 1981. *Forest Island Dynamics in Man-Dominated Landscapes*. Springer, New York.

Caffey, H.M. 1985. Spatial and temporal variation in settlement and recruitment of intertidal barnacles. *Ecol. Monogr.* 55(3):313–332.

Campbell, J.B. 1977. Variation of selected properties across a soil boundary. *Soil Sci. Soc. Am. J.* 41:578–582.

Carter, D.B., Schmudde, T.H., Sharpe, D.M. 1972. The interface as a working environment: A purpose for physical geography. *Technical Paper No. 7* Association of American Geographers Commission on College Geography, Washington, D.C.

Casetti, E. 1970. Generating models by the expansion method: Applications to geographical research. Paper presented at IGU Quantitative Commission Conference, Poznan, Poland.

Clark, J. 1976. *The Sanibel Report*. Conservation Foundation, Washington.

Cliff, A.D., Ord, J.K. 1973. *Spatial Autocorrelation*. Pion, London.

Cooper, J.P. (ed.) 1975. *Photosynthesis and Productivity in Different Environments*. IBP series, Vol. 3. Cambridge University Press, Cambridge.

DeAngelis, D.L., Waterhouse, J.C., Post, W.M., O'Neill, R.V. 1985. Ecological modeling and disturbance evolution. *Ecol. Model.* 29:339–419.

Dahiya, I.S., Anlauf, R., Kersebaum, K.C., Richter, J. 1985. Spatial variability of some nutrient constituents of an Alfisol from loess: 2 Geostatistical analyses. *Z. Pflanzenernaehr. Bodenkd.* 148(3):268–277.

Delcourt, H.R., Delcourt, P.A., Webb, T. 1983. Dynamic plant ecology: The spectrum of vegetational change in space and time. *Quat. Sci. Rev.* 1:153–175.

Dickerman, J.A., Stewart, A.J., Wetzel, R.G. 1986. Estimates of net annual aboveground production: Sensitivity to sampling frequency. *Ecology* 67:650–659.

Dorney, R.S., McClellan, P.W. 1984. The urban ecosystem: Its spatial structure, its scale relationships, and its subsystem attributes. *Environments* 16:9–20.

Dyer, M.L. 1986. A model of organic decomposition rates based on climate and litter properties. Master's thesis, University of Georgia, Athens, GA.

Edmonds, W.J., Baker, J.C., Simpson, T.W. 1985. Variance and scale influences on classifying and interpreting soil map units. *Soil Sci. Soc. Am. J.* 49(4):957–961.

Flenley, J.R. 1984. Time scales in biogeography, pp. 63–105. *In* J.A. Taylor (ed.), *Themes in Biogeography*. Croom Helm, London.

Folkoff, M.E., Meentemeyer, V., Box, E.O. 1981. Climatic control of soil acidity. *Phys. Geogr.* 2:116–124.

Forman, R.T.T., Godron, M. 1986. *Landscape Ecology*. Wiley & Sons, New York.

Galiano, E.F., Sterling, A., Viejo, J.L. 1985. The role of riparian forests in the conservation of butterflies in a Mediterranean area. *Environ. Conserv.* 12(4):361–362.

Godron, M., Forman, R.T.T. 1983. Landscape modifications and changing ecological characteristics, pp. 12–28. *In* H.A. Mooney and M. Godron (eds.), *Disturbance and Ecosystems: Components of Response*. Springer-Verlag, New York.

Haggett, P., Chorley, R.J., Stoddart, D.R. 1965. Scale standards in geographical research: A new measure of area magnitude. *Nature* 205:884–847.

Haggett, P., Cliff, A.D., Frey, A. 1977. *Locational Analysis in Human Geography*. Edition 2. Wiley & Sons, New York.

Hansen, J., Russell, A., Lacis, A., Fung, I., Rind, D., Stone, P. 1985. Climate response times: Dependence on climate sensitivity and ocean mixing. *Science* 229(4716):857–859.

Hardin, G., Baden, J. 1977. *Managing the Commons*. Freeman, San Francisco.

Harris, L.D. 1984. *The Fragmented Forest: Island Biogeography Theory and the Preservation of Biotic Diversity*. University of Chicago Press, Chicago.

Hesketh, J.D., Jones, J.W. (eds.) 1980. *Predicting Photosynthesis for Ecosystem Models*. 2 Vols. CRC Press, Boca Raton, FL, pp. 273–279.

Hill, M.O. 1973. The intensity of spatial patterns in plant communities. *J. Ecol.* 61:225–235.

Hoover, S.R. 1986. Comparative landscape structure across physiographic regions of Georgia. Master's thesis, University of Georgia, Athens, GA.

Imeson, A.C. 1984. Geomorphological processes, soil structure, and ecology, pp. 72–84. *In* A. Pitly (ed.), *Geomorphology: Themes and Trends*, Barnes and Noble Books, Totowa, NJ.

Johnson, W.C., Sharpe, D.M. 1976. An analysis of forest dynamics in the northern Georgia piedmont. *Forest Sci.* 22:307–322.

Jones, D.W., Krammel, J.R. 1985. The location theory of animal populations: The case of a spatially uniform food distribution. *Am. Nat.* 126(3):392–404.

Kershaw, K.A. 1970. An empirical approach to the estimation of pattern intensity from density and cover data. *Ecology* 51:729–734.

Kesner, B.T. 1984. The Geography of Nitrogen in an Agricultural Watershed: A Technique for Spatial Accounting of Nutrient Dynamics. Master's thesis, University of Georgia, Athens, GA.

Landsberg, H.E. 1983. Variability of the precipitation process in time and space, pp. 3–9. *In* S.A. Campbell (ed.), *Sampling and Analysis of Rain. ASTM STP 823*. Society for Testing and Materials, Philadelphia.

Lieth, H. (ed.) 1974. *Phenology and Seasonality Modeling*. Ecological Studies, Vol. 8. Springer-Verlag, New York.

Lieth, H. 1975. Modeling the primary productivity of the world, pp. 237–263. *In* H. Lieth and R.H. Whittaker (eds.), *Primary Productivity of the Biosphere*. Springer-Verlag, New York.

Lewin, R. 1985. Plant communities resist climatic change. *Science* 228:165–166.

Margalef, R. 1968. *Perspectives in Ecological Theory*. University Chicago Press, Chicago.

Matheron, G. 1971. *The Theory of Regionalized Variables and its Applications*. Ecole de Mines, Fontainbleau, France.

McBratney, A.B., Webster, R., McLaren, R.G., Spiers, R.B. 1982. Regional variation of extractable copper and cobalt in the topsoil of southeast Scotland. *Agronomics* 2(10):962–982.

McHarg, I.L. 1971. *Design With Nature*. Doubleday/Natural History Press, Garden City, NY.

Meentemeyer, V. 1978. Macroclimate and lignin control of litter decomposition rates. *Ecology* 59:465–472.

Meentemeyer, V., Box, E.O., Thompson, R. 1982. World patterns and amounts of terrestrial plant litter production. *BioSci.* 32:125–128.

Meentemeyer, V. 1984. The geography of organic decomposition rates. *Ann. Assn. Am. Geogr.* 74:551–560.

Meentemeyer, V., Gardner, J., Box, E.O. 1985. World patterns and amounts of detrital soil carbon. *Earth Surf. Process. Landforms* 10:557–567.

Michener, W.K. (ed.) 1986. *Research Data Management in the Ecological Sciences*. University of South Carolina Press, Columbia.

Miller, D.H. 1978. The factor of scale: Ecosystem, landscape mosaic, and region, pp. 63–88. *In* K.A. Hammond (ed.), *Sourcebook on the Environment*. University of Chicago Press, Chicago.

Mooney, H.A., Godron, M. (eds.) 1983. *Disturbance and Ecosystems: Components of Response*. Springer-Verlag, New York.

Muehrcke, P.C. 1980. *Map Use: Reading, Analysis and Interpretation*. J.P. Publications, Madison, WI.

NASA 1986. *Earth System Science Overview: A Program for Global Change*. Earth System Science Committee, NASA Advisory Council. National Aeronautics and Space Administration, Washington, D.C.

Naveh, Z., Lieberman, A.S. 1984. *Landscape Ecology: Theory and Applications*. Springer-Verlag, New York.

Noss, R.F. 1983. A regional landscape approach to maintain diversity. *BioSci.* 33:700–706.

Ophuls, W. 1977. *Ecology and the Politics of Scarcity*. Freeman, San Francisco.

Peters, R.H. 1983. *The Ecological Implications of Body Size*. Cambridge University Press, Cambridge.

Phipps, M. 1981. *Information Theory and Landscape Analysis*. Proceedings of the International Congress of the Netherlands Society of Landscape Ecology, Veldhoven, Wageningen, Pudoc.

Pickett, S.T.H., White, P.S. (eds.) 1985. *The Ecology of Natural Disturbance and Patch Dynamics*. Academic Press, Orlando, FL.

Pielou, E.C. 1975. *Ecological Diversity*. Wiley-Interscience, New York.

Plowright, R.C., Galen, C. 1985. Landmarks of obstacles: The effects of spatial heterogeneity on bumble bee (Bombus) foraging behavior. *Oikos* 44(3):459–465.

Reiners, W.A. 1983. Disturbance and basic properties of ecosystem energetics, pp. 83–98. *In* H.A. Mooney and M. Godron, (eds.), *Disturbance and Ecosystems. Components of Response*. Springer, New York.

Risser, P.G., Karr, J.R., Forman, R.T.T. 1983. *Landscape Ecology: Directions and Approaches*. IL. Nat. Hist. Survey Spec., Publ. No. 2.

Roels, J.M. 1985. Estimation of soil loss at a regional scale based on plot measurements—some critical considerations. *Earth Surf. Process Landforms* 10:587–595.

Romme, W.H., Knight, D.H. 1982. Landscape diversity: The concept applied to Yellowstone Park. *BioSci.* 32:664–670.

Salthe, S.N. 1985. *Evolving Hierarchical Systems: Their Structure and Representation.* Columbia University Press, New York.

Schneider, D.C., Duffy, D.C. 1985. Scale-dependent variability in seabird abundance. *Mar. Ecol. Prog. Ser.* 25(3):211–218.

Schreiber, K.F. 1986. What is Landscape Ecology? International Association for Landscape Ecology, Wageningen. *IALE Bulletin* 4(1):8–13.

Sharpe, D.M., Stearns, F.W., Burgess, R.L., Johnson, W.C. 1981. Spatio-temporal patterns of forest ecosystems in man-dominated landscapes of the eastern United States, pp. 109–116. *In* S.P. Tjallingii and A.A. de Veer (eds.), *Perspectives in Landscape Ecology.* Centre for Agricultural Publication and Documentation, Wageningen.

Shepard, P., McKinley, D. (eds.) 1969. *The Subversive Science: Essays Toward an Ecology of Man.* Houghton-Mifflin, Boston, MA.

Shugart, H.H., Seagle, S.W. 1985. Modeling forest landscapes and the role of disturbance in ecosystems and communities, pp. 351–384. *In* S.T.A. Pickett and P.S. White (eds.), *The Ecology of Natural Disturbance and Patch Dynamics.* Academic Press, Orlando, FL.

Steele, J.H. (ed.) 1977. *Spatial Pattern in Plankton Communities.* Plenum Press, New York and London.

Thornthwaite, C.W., Mather, J.R. 1955. *The Water Balance.* Public. Climatol., 8(1):1–104.

Tjallingii, S.P., de Veer, A.A. (eds.) 1981. *Perspectives in Landscape Ecology.* Centre for Agricultural Publication and Documentation (PUDOC), Wageningen.

Tobler, W.R. 1969. Geographical filters and their inverses. *Geographical Analysis* 1:234–253.

Tomlin, C.D. 1980. *The Map Analysis Package.* Yale University School of Forestry and Environmental Studies, New Haven, CT.

Troll, C. 1939. Luftbildplan und ökologische Bodenforschung. *Z. Ges. Erdkunde (Berlin)* pp. 241–298.

Troll, C. 1950. Die geographische Landschaft und ihre Erforschung. *Studium Generale (Heidelberg)* 3:163–181.

Troll, C. 1966. *Landscape Ecology.* Publ. S-4. ITC-UNESCO Centre, Delft.

Troll, C. 1968. Landschaftsökologie, pp. 1–21. *In* R. Tüxen (ed.), *Pflanzensoziologie und Landschaftsökogologie.* Dr. W. Junk bv, The Hague.

Tucker, C.J., Townshend, J.R.G., Goff, T.E. 1985. African landcover classification using satellite data. *Science* 227:369–376.

Tüxen, R. (ed.) 1968. *Pflanzensoziologie und Landschaftsökologie.* Dr. W. Junk bv, The Hague.

Vink, A.P.A. 1983. *Landscape Ecology and Land Use.* Longman, London and New York.

Watts, D. 1984. The spatial dimension in biogeography, pp. 25–62. *In* J.A. Taylor (ed.), *Biogeography: Recent Advances and Future Directions.* Barnes and Noble Books, Totowa, NJ.

Weinberg, G.M. 1975. *An Introduction to General Systems Thinking.* Wiley & Sons, New York.

White, P.S. 1979. Pattern, process, and natural disturbance in vegetation. *Bot. Rev.* 45:229–299.

Wickham, J.D. 1984. Climatic Correlates of the Initiation of the Growing Season for the Eastern United States, for the Spring Period 1984. Master's thesis, University of Georgia, Athens, GA.

Wiens, J.A. 1976. Population responses to patchy environments. *Ann. Rev. Ecol. Syst.* 7:81–120.

Wiens, J.A., Crawford, C.S., Gosz, J.R. 1985. Boundary dynamics: A conceptual framework for studying landscape ecosystems. *Oikos* 45:421–427.

Wiesner, C.J. 1970. *Hydrometeorology*. Chapman and Hall, London.

Wojcik, J., Meentemeyer, V., Box, E.O. (1987). Climatic control of base saturation in United States soils. *Soil Sci.* In press.

Woodmansee, R.G., Adamsen, F.J. 1983. Biogeochemical cycles and ecological hierarchies, pp. 497–516. *In* R.R. Lowrance, R.L. Todd, L.E. Asmussen and R.A. Leonard (eds.), *Nutrient Cycling in Agricultural Ecosystems*. Special Publ. No. 23. University of Georgia College of Agricultural Extension Stations, Athens, GA.

2. Disturbances in the Landscape

3. Landscape Ecology and Air Pollution*

F.H. Bormann

3.1 Objectives

Much of the work of landscape ecologists starts from the notion that landscape is a collection of discrete sites and that by proper planning and management these sites can accomplish important societal tasks, continue to perform vital natural functions, and in general enhance the enjoyment and dignity of present and future generations.

Although these objectives remain important, a new and significant threat to the health of natural and managed landscapes has emerged. It is based on the fact that all parts of the world are bound together by gigantic natural pathways of air and water, pathways that ignore human-made boundaries, and that the human's capacity to contaminate these pathways has reached appalling proportions and gives every indication of still greater contamination to come. I suggest that the efforts of landscape managers to understand pathway contamination and its effects on natural and managed ecosystems and to work for control are, over the long term, just as important as are efforts to manage and preserve individual units or classes of units.

In this chapter I examine the question of regional air pollution and its potential effects on managed and natural ecosystems, particularly forest ecosystems. To

*Adapted from Bormann, F.H. 1985. Air pollution and forests: An ecosystem perspective. *BioScience* 35:434–441.

do so, I briefly discuss the structure and function of ecosystems, present a model of forest ecosystem response to air pollution stress, and present some evidence on regional air pollution levels in North America and some data linking air pollution deposition with damage to ecosystems. I conclude with a brief discussion of how landscape heterogeneity and ecosystem complexity make it exceedingly difficult to link air pollution deposition in a quantitative way to ecosystem damage. Nevertheless, circumstantial evidence and ecological models of the near future indicate a strong necessity for more stringent emission controls.

3.2   Linkages and Biotic Regulation

To discuss the effect of regional air pollution stress on natural and managed ecosystems, it is necessary to consider energy relationships of these systems and the linkage of forest ecosystems to interconnected ecosystems and to climate. Forest ecosystems are driven directly and indirectly by solar energy. The quantities passing through forest ecosystems are huge. For example, in New England, natural ecosystems use about 87 quads per year or 30 times that used by humans (Bormann 1982). Solar energy not only drives ecosystem processes such as photosynthesis, transpiration, nutrient uptake, mineralization, nitrogen fixation, and growth of all organisms, but considerable quantities are stored in dead organic matter, which has important regulatory functions within the ecosystem.

Research at Hubbard Brook, Coweeta, H.J. Andrews Experimental Forests, the Institute of Ecology, and elsewhere have shown how, through the direct and indirect use of solar energy (Fig. 3.1), the terrestrial ecosystem carries out a number of processes that determine (1) the form in which solar energy entering the system is stored and/or returned to the atmosphere, (2) the chemical makeup of the flow of air and water moving through the ecosystem boundaries, (3) the proportion of water that will leave as a liquid or as vapor, (4) the proportion of liquid water that leaves as surface runoff or seeps into the ground to recharge ground water, and (5) the amount of material eroded from the ecosystem.

The sum of these activities may be thought of as biotic regulation of the flow of energy, water, and chemicals through the ecosystem and regulation of outputs from the ecosystem (Likens et al. 1977; Bormann and Likens 1979a).

Through biotic regulation the terrestrial ecosystem governs, to some degree, the behavior of interconnected stream and lake ecosystems (Likens and Bormann 1974; Bormann and Likens 1985), local and regional climate (Reifsnyder 1984), and may even affect the global carbon cycle by net storage or release of carbon dioxide (Woodwell et al. 1978). This means that effects of air pollution stress may not be limited to the individual ecosystem but may be transmitted to interconnected systems by air and water pathways.

3.3   A Model of Ecosystem Decline

Studies of forest ecosystems, around strong point sources of air pollution and radiation pollution by Gordon and Gorham (1963), Woodwell (1970), Knabe

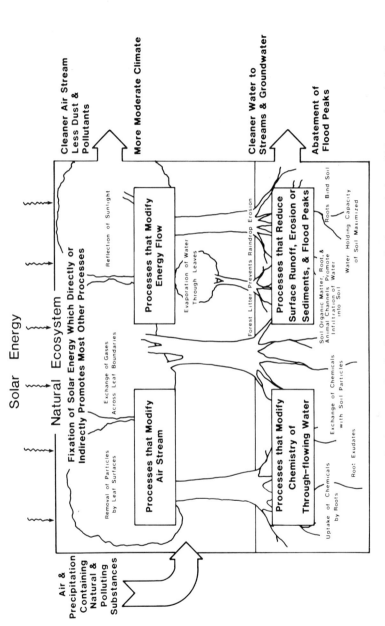

Figure 3.1. Some of the uses of solar energy by the forest ecosystem in the regulation of biogeochemical cycles.

(1976), and others, and results from laboratory and field experiments summarized by Smith (1981) suggest that forest ecosystems respond to increasingly severe pollution in a repeatable pattern.

The pattern of decline under increasing stress caused by air pollution can be thought of as a continuum of stages leading to ecosystem collapse (Bormann 1982; Fig. 3.2).

Stage 0: Anthropogenic pollutant levels insignificant. Pristine systems.

Stage I: Anthropogenic pollution occurs at generally low levels. Ecosystems serve as a sink for some pollutants, but species and ecosystem functions are relatively unaffected.

Stage IIA: Levels of pollutants are inimical to some aspect of the life cycle of sensitive species or individuals. As a result they are subtly and adversely affected. For example, sensitive plants may have reduced photosynthesis, a change in reproductive capacity, or a change in predisposition to insect or fungus attack.

Stage IIB: With increased pollution stress, populations of sensitive species decline and their effectiveness as functional members of the ecosystem diminishes. Ultimately, these species may be lost from the system, but a more likely fate is that some individuals will remain as insignificant components.

Stage IIIA: With still more pollution stress, due to higher concentrations or longer exposure at the same concentration, the size of the plant becomes important to survival, and large plants, trees, and shrubs of all species die. The

Model of Ecosystem Collapse

Stage	Level of Anthropogenic Pollution	Severity of Impact on Ecosystem
0	Insignificant	Insignificant
I	Low level	Relatively unaffected; may serve as sink
IIA	Levels inimical to sensitive organisms	Changes in competitive ability of sensitive species; selection of resistant genotypes; little effect on biotic regulation
IIB	Increased pollution stress	Resistant species substitute for sensitive ones; some niches opened for lack of substitutes; biotic regulation may be disrupted, but may return as system becomes wholly populated by resistant species
IIIA	Severe levels of pollution	Large plants, trees, shrubs of all species die off; ecosystem converted to open-small shrubs, weedy herb system; biotic regulation severely diminished; increased runoff, erosion, nutrient loss
IIIB	Continued severe pollution	Ecosystem collapse; completely degraded ecosystem; ecosystem seeks lower level of stability with much less control over energy flow and little biotic regulation

Figure 3.2. Model for ecosystem decline under air pollution stress.

basic structure of the forest ecosystem is changed (Woodwell 1970), and biotic regulation is affected. Both Gordon and Gorham (1963) and Woodwell (1970) describe the process as peeling off layers of forest structure: first the trees, followed by tall shrubs, and finally under the severest conditions, the short shrubs and herbs. The structure of the forest ecosystem is progressively changed to one dominated by small scattered shrubs and herbs, including weedy species not previously present in the forest. Productivity is greatly reduced as the ability of the ecosystem to repair itself by substituting tolerant for intolerant species is exceeded. As layers of vegetation die, masses of highly flammable dead wood are left behind, increasing the probability that natural or human-induced fire will occur and that conversion to open shrub land will be speeded. Toxic concentrations of some pollutants may limit many species (Jordan 1975; Hutchinson 1980). The capacity of the ecosystem to regulate energy flow and biogeochemical cycles is severely diminished. Runoff is increased; the loss of nutrients previously held and recycled is accelerated; erosion is increased and soil and nutrients are exported to interconnected aquatic systems, which may be severely affected (Bormann et al. 1974; Likens 1984).

Stage IIIB: Ecosystem collapse. The ecosystem at this point is so damaged by loss of species, ecosystem structure, nutrients, and soil that the capacity to repair itself is severely diminished. Even if the perturbing force is removed predisturbance levels of structure and function may never be achieved or, if they are, would take centuries or even millennia to accomplish. This type of degraded ecosystem is frequently seen around strong point sources of air pollution active in pre-emission-control days, such as Copper Hill, Tennessee and Sudbury, Ontario.

Disturbance in nature is not unusual and all ecosystems are subject to various kinds of natural or human-made recurring disturbances, such as wild fires, wind storms, insect outbreaks, or forest cutting. Many ecosystems have a built-in capacity to recover from these disturbances (Bormann and Likens 1979a,b). After the perturbing event is over in the course of decades, the system will rebuild itself to predisturbance levels of structure and function. The difference between the classical disturbances discussed above and stage III stress resulting from air pollution is the long-term nature of the disturbance and the degree of degradation, which drive the ecosystem over the threshold of resilience where the properties of the system may seek new steady state levels instead of returning to original conditions (Margalef 1969; Hill 1975; Bormann and Likens 1979a).

3.4 Redundant Functions of Ecosystems

We now need to consider an important property of some ecosystems whereby they can perform important functions in more than one way or have a reserve capacity to perform important functions beyond current needs; we call this collective property redundancy. The reserve capacity aspect of redundancy is based in part on abiotic features such as soil structure, which in some systems can for a time resist soil erosion even though the plants of the system have been destroyed (Bormann and Likens 1979a). For example, at Hubbard Brook, we deforested an entire watershed by cutting the forest and leaving the fallen trees. Regrowth was prevented for three years by herbicide application. The forest floor was able to control surface erosion for two years, despite the fact that its renewal through litter fall was cut off by the treatment and water flows were greatly increased (Bormann et al. 1974). Redundancy is also based on substitution or the ability of the ecosystem to accomplish vital functions such as photosynthesis, decomposition, nutrient uptake, nutrient storage, and water routing in more than one way. Thus, if one pathway is impaired, another pathway may take over with little long-term change in the ecosystem's capacity to fix energy and to achieve biotic regulation. The substitution aspect of redundancy is founded on variations among genotypes, populations, and species which allow the ecosystem flexibility in "adapting" to stress (Bradshaw 1976). At the community level, redundancy is based on having more than one species of plant, animal, or microbe capable of carrying out the same process such as photosynthesis, pollination, seed dispersal, and nitrification. One or several species may substitute for an impaired species and assume most of its functions.

As our model of ecosystem stress indicates, redundancy can operate to diminish loss of biotic regulation. Thus, in stage II, reserve capacity, genotypic and species substitutions, and use of alternative pathways can for a time maintain the ecosystem's integrity in the face of air pollution stress. But there is a price to pay; as redundancy is used up, the ecosystem becomes increasingly vulnerable to added stress and to a stage III response.

3.5 Regional Stress on Forest Ecosystems

There are numerous examples and substantial agreement about the effects of air pollution stress around strong point sources, but potentially damaging levels of air pollution are occurring over huge areas.

Increased acidity of precipitation now occurs over large areas of North America (Fig. 3.3), Europe, and China (Zhao 1985). Galloway et al. (1984) estimate that concentrations of sulfate, a significant indicator of anthropogenic air pollution, are enriched 2 to 16 times in precipitation falling on eastern North America relative to that falling on remote areas. Ozone at concentrations known to cause plant damage (National Academy of Science 1977; Environmental Protection Agency 1984a) occurs over wide areas of the United States (Fig. 3.4). Even so, the full magnitude of the ozone or more generally, the photochemical oxidant problem is unknown because very few monitors are located in rural areas.

A preliminary study during 1981 to 1982 indicated that cloud moisture being deposited on vegetation is more acid than rain and has a higher concentration of associated chemicals (Table 3.1). These differences are not insignificant. Concentrations of hydrogen ions and associated anions and cations in cloud water can exceed those in rain by several times.

In 1983, Likens and I with cooperators established the North American Cloud

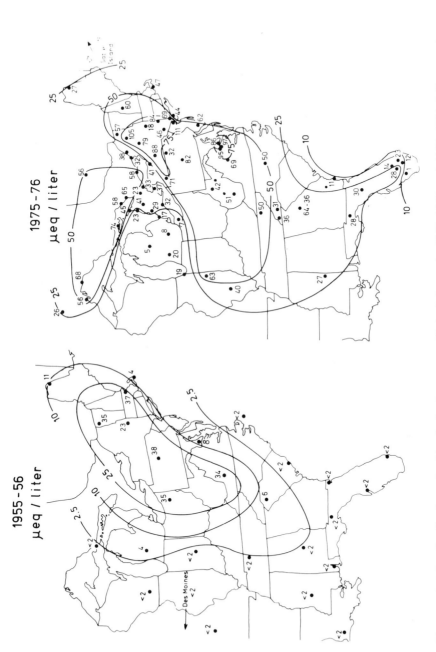

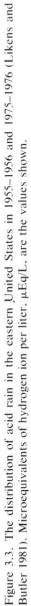

Figure 3.3. The distribution of acid rain in the eastern United States in 1955–1956 and 1975–1976 (Likens and Butler 1981). Microequivalents of hydrogen ion per liter, µEq/L, are the values shown.

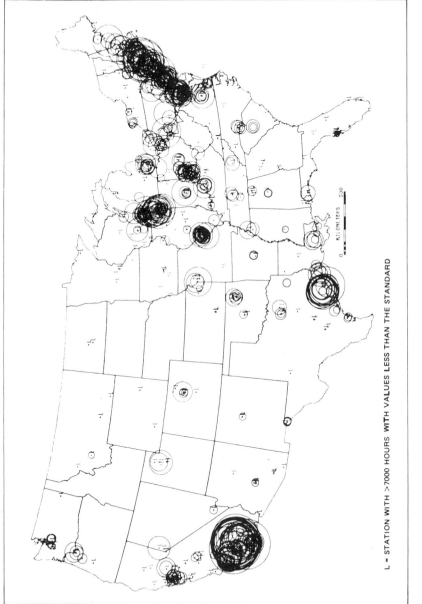

Figure 3.4. Areas where it is 95% probable that the NAAQS for ozone (0.12 ppm) was exceeded during 1977 and 1978 (Ludwig and Shelar 1980).

Table 3.1. Preliminary Data on Differences Between Chemistry of Cloud Water and Bulk Precipitation on Mounts Washington and Lafayette, New Hampshire[a]

Item	Clouds	Bulk Precipitation	Clouds/Bulk Precipitation
pH	3.68	4.11	2.69[b]
Cl	0.51[c]	0.68	0.8
Na	0.35	0.13	2.7
SO_4	11.54	3.71	3.1
NH_4	1.68	0.51	3.3
Ca	0.99	0.30	3.3
K	0.22	0.05	4.4
NO_3	7.06	1.43	4.9
PO_4	0.03	0.006	5.0
Mg	0.26	0.04	6.5

[a] Collected at approximately 1500 m. Data pooled from summers of 1981 and 1982. Sample size for clouds, 36; for bulk precipitation, 30 (Bormann 1983).
[b] Based on H^+ concentrations.
[c] Amounts given in milligrams per liter.

Moisture project. Results from this study not only illustrate another aspect of the acid rain problem but also the difficulty of considering the effects of pollutants separately. One cloud event recorded during August 7 to 12, 1984, apparently covered thousands of square miles in the northeast and deposited cloud moisture on vegetation ranging from pH 2.80 to 2.97 (Weathers et al. 1986). This is about 20 times more acidic than the average acidity (pH 4.1) of rain falling in the region. On August 7, an ozone monitor operated by the Department of Environmental Protection (DEP), State of Maine, a 5-km distance from our cloud moisture station on Mount Desert Island, Maine, recorded six hours of hourly means exceeding 0.09 ppm (Environmental Protection Agency 1984b). This level is known to be damaging to many species of plants (Environmental Protection Agency 1984a). It was also on August 7 that our monitor recorded cloud moisture of pH 2.8. Thus, vegetation on Mount Desert Island was subjected, more or less simultaneously, to extremes of two well-known stresses.

Atmospheric fallout of trace metals, potentially toxic to organisms in small amounts, have greatly increased as a result of industrial activities (Galloway et al. 1982). Lead, zinc, manganese, silver, arsenic, vanadium, antimony, selenium, chromium, and nickel have been estimated to be 10 to 200 times more concentrated in rural continental regions than in remote areas such as the South Pole. Rural values are not always on the trivial end of the scale. Lead concentrations in the forest floor of Camel's Hump Mountain in Vermont, far from emission sources, were equivalent to high values measured in forests near urban areas (Andresen et al. 1980). It seems reasonable to assume that regional air pollution is often a variable mix of acid rain, photochemical oxidants, trace metals, and other substances (Last 1984).

The complexity of estimating air pollution deposition in rural locations can be seen in a photomosaic of the eastern United States taken from 540 miles in space at night (Fig. 3.5). Each patch of light is a center of fossil and nuclear

Figure 3.5. A satellite photomosaic of eastern United States and a small portion of southeastern Canada roughly bounded by 25° to 42° latitude and the eastern seaboard to about 97°W longitude. This photograph was taken at night from an elevation of 540 nautical miles. The white areas are the lights of urban areas and represent centers of energy consumption. This area is largely contained within the boundaries of the temperate eastern deciduous forest. Note that no area of the eastern forests is very far removed from an urban-industrial center (Bormann 1976).

energy consumption from which a variety of air pollutants radiate outward into surrounding natural ecosystems. No part of the eastern deciduous forest is very far from a center. In many areas throughout the world, emissions from centers like these complicate deposition patterns at various distances from their periphery.

3.6   Subtle and Long-Term Effects of Regional Pollutants

These deposition patterns suggest that stage II responses may be occurring over large areas, and although difficult to detect, they may have subtle, cumulative, and potentially damaging long-term implications.

One such implication is the possibility of significant changes in gene pools (Antonovics 1968; Houston and Stairs 1973; Bradshaw 1976; Karnosky 1977; Ayazloo and Bell 1981) in areas that for a long time have had chronic pollution at levels capable of producing stage II responses. For example, are species undergoing genetic change with natural selection favoring resistance to air pollution? Does acquisition of genetically based resistance to air pollution have hidden costs like lowered productivity and loss of fitness (Bradshaw 1976); for example, loss of competitive ability; loss of reproductive capacity; or resistance to drought, insects, or disease. Are whole regions, such as Southern Connecticut and adjacent New York where there has been 30 years of photochemical-oxidant pollution, undergoing significant contractions of gene pools.

There is some evidence that air pollution stress is affecting productivity of regional terrestrial ecosystems (Bormann 1982). Extensive field and laboratory work has documented that many plant species are sensitive to sulfur dioxide and/or ozone, often at concentrations below U.S. air quality standards (Davis and Wilhour 1976; Manning 1979; Williams and Wong 1980; EPA 1984a). Tree ring analysis in fairly remote places in California (Miller et al. 1976) and Virginia (Skelly 1980; Williams 1980; Bennoit et al. 1982) suggest declines in productivity associated with symptoms of photochemical-oxidant damage. Symptoms indicative of photochemical-oxidant damage on plants have been recorded in more than 28 states (EPA 1984a; Fig. 3.6) and have been noted on trees in widespread areas in Indiana, Wisconsin, New England and North Carolina (personal communications: Wayne T. Swank, Coweeta Hydrologic Laboratory, Otto, NC;

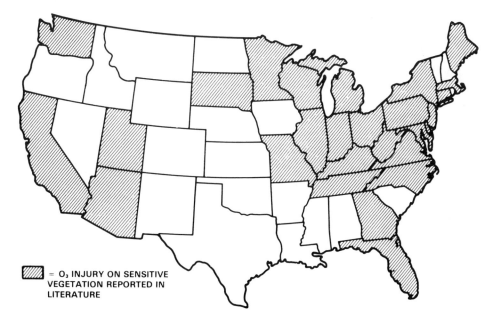

= O₃ INJURY ON SENSITIVE VEGETATION REPORTED IN LITERATURE

Figure 3.6. States in which some injury to vegetation has occurred as reported in the published literature (EPA 1984a).

and Michael Treshow, University of Utah, Salt Lake City, October 1984; and Neagele et al. 1972, Usher and Williams 1980; EPA 1984b).

Open-top chamber studies have reported yield declines in crops due to air pollution in England (Bell and Clough 1973) and productivity declines in native grass populations in remote areas of western Virginia (Duchelle et al. 1983). Field experiments with soybeans in Illinois have indicated that productivity losses occurred without visible symptoms of damage (Sprugel et al. 1980). In perhaps the most comprehensive study, the National Crop Loss Assessment Network, using open-top chambers and controlled levels of ozone under field conditions, has estimated regional productivity losses in crop plants due to ozone at ambient concentrations. Test plots in North Carolina, New York, Illinois, and California have detected yield losses of 15% to greater than 50% on crops common to each region (personal communication, Leonard Weinstein, Boyce Thompson Institute, Cornell University, Ithaca, NY, December 1984; Heck et al. 1982). In our work at the Institute of Ecosystem Studies, Wang et al. (1986) demonstrated a 15 to 20% reduction in aspen, eastern cottonwood, and hybrid poplar productivity under ambient ozone conditions without visible symptoms of ozone damage.

Collectively, these data raise the possibility that presently air pollution stress is causing potentially significant but largely unmeasurable declines (to be further explained later) in net primary productivity or energy flow of forest and field ecosystems. Such declines in productivity may not be now associated with loss of biotic regulation. As populations of sensitive plant species decline under air pollution stress, their role in ecosystem function may be taken over by tolerant species in the same or different layers of the forest. Tolerant species may prosper because of release from competition with sensitive species for available resources within the ecosystem. These population changes may be enormously significant as economically important species are replaced by less important species, or they may go unnoticed if the species being lost are not of economic importance. However, in both instances, biotic regulation of energy flow, hydrology, erosion, gaseous exchange or nutrient loss may be little affected by changes in species importance.

The situation just described postulates a threshold below which biotic regulation by the ecosystem may not be affected by pollution effects on internal components of the ecosystem. In fact the above situation may be more analogous to the finding of radiation biologists about two decades ago that biological effects of radiation were of a continuous nature rather than a previously conceived threshold effect. Thus, small but continuous reductions in energy flow in ecosystems may be linked to subtle declines in the ecosystem's capacity to regulate energy flow and biogeochemical cycles. Because photosynthetic fixation of solar energy by plants is the cornerstone of all ecosystem activities, chronic depression of photosynthesis raises questions of fundamental importance about indirect effects of air pollution on ecosystem function. What would happen for example if net productivity of natural ecosystems were reduced 10% because of chronic air pollution? Would reduced net carbon dioxide fixation over wide geographic areas contribute significantly to atmospheric carbon dioxide concentration?

Would more runoff and a greater loss of nutrients in drainage water eventually result? Would the rate of erosion subtly increase as productivity declines? Would species diversity be lowered in relation to reductions in energy flow? Would the ecosystem undergo creeping degradation largely imperceptible because of its slow rate?

3.7 Environmental Decision Making in the Face of Scientific Uncertainty

My comments on genetic and productivity responses of ecosystems are based largely on limited but important experimental studies. However, some economists, politicians, and industrialists are asking for hard field data that quantitatively link forest damage to emissions of specific air pollutants. They ask, "What will billions of dollars spent in emission controls achieve in restoring economic benefits derived from natural ecosystems?" A corollary is that we should move slowly in imposing new emission controls until the cost/benefit relationship is quantified. This position is often coupled with a call for more research before action is taken (U.S. Congress 1982). Although this attitude may be understandable in persons whose primary concerns are economic and whose knowledge and concern about environmental problems is limited, this is one of those cases described by Grobstein (1983) where a decision cannot be made with scientific certainty but must be made with the best available scientific information.

The quantification of the dollar cost of air pollution stress on forest and other naturally occurring ecosystems at places distant from emission sources is proving to be exceedingly difficult if not impossible. *Landscape heterogeneity* plays an important role in this calculation.

There are places where regional ecosystem responses similar to stage IIB in our model are measurable, such as in the Federal Republic of Germany where there is good quantitative data documenting the crash of *Picea abies* and the decline of *Abies alba* under a variety of circumstances (Anonymous 1982), and in the northeastern United States where there are quantitative data documenting the crash of *Picea rubra* and the potential decline of *Abies balsamea* during the last two decades (Fig. 3.7; Johnson and Siccama 1983). In these areas scientists are attempting to identify mechanisms and processes causing decline in these species and to determine whether air pollution is a causative factor (Ulrich and Pankrath 1983), but so far the explicit connection of air pollution to these ecosystem responses has yet to be established (Johnson and Siccama 1983; Burgess 1984; Society of American Foresters 1984). In California where *Pinus ponderosa* and *Pinus jeffryi* are in decline over extensive areas, the qualitative connection of the decline with ozone is well established, but little is known of the quantitative importance of ozone and the other pollutant components of the smog affecting the area, climatic variables, or insects and pathogens (Miller et al. 1976; Williams and Williams 1984).

Figure 3.7. Camel's Hump Mountain, Vermont, showing forest damage. In 1964 numerous *Picea rubens*, 200 to 300 years old, were found on the west slope. Density of trees ≥ 10 cm diameter at breast height declined 52% and basal area declined 44% between 1965 and 1979. Density and basal area of standing dead spruce in 1979 exceeded that of living stems (Siccama et al. 1982).

This is our dilemma. If we cannot be positive about heavily studied type IIB responses, how can we hope to measure and quantitatively evaluate, under field conditions, the much more subtle and potentially more widespread type IIA responses? Thus, even though terrestrial ecosystems in large regions may be, and probably are, responding to air pollution stress, we do not have as yet adequate means for measuring these responses. This is not surprising given that (1) our knowledge of actual pollution inputs is fragmentary, (2) long-term ecosystem data are rare, and (3) even if growth or population trends are detected, they are difficult to relate in a quantitative way to specific air pollutants.

To link forest damage in a quantitative way to air pollution, we need to know what kinds and how much pollution is entering the ecosystem under study. Not only is regional air pollution a variable mixture of acids, photochemical oxidants, trace metals, synthetic organics, and gases, but local deposition may depart significantly from deposition estimates made on the basis of regional means. Due to landscape heterogeneity, some local sites may receive more intense deposition due to orographic rain, wind exposure, vegetation type, or other factors, whereas nearby sites receive less deposition than the regional mean because of rain shadows, wind protection, or vegetation type. The land-scape is probably best considered as a varying patchwork of deposition governed by form of pollutant, mode of deposition, vegetation structure of the site, to-pography, and local meteorology (Bormann 1983).

Another aspect of quantifying air pollution damage to ecosystems is quan-tification of changes in the structure and function of the ecosystem under stress. To do this it is necessary to have long-term data to provide a standard against which we can measure changes or trends in population composition, ecosystem structure, or biotic regulation attributable to changes in air quality (Likens 1983). Such data are rare because funding for long-term research is scarce and such research is difficult.

Even when trends are detected, such as through tree-ring analysis, relating these trends to air pollution is difficult not only due to uncertainty about air pollution input but also due to background noise present in almost any eco-system. Background variation may be caused by:

1. fluctuations in variables controlling growth and productivity such as pre-cipitation, temperature, length of growing season, frosts, and drought
2. developmental processes such as succession
3. stochastic events like insect outbreaks, disease epidemics, fire, cutting, or windstorm (Bormann 1983, Burgess 1984).

To assign a causation value for any identified trend, the investigator must be able to partition causation among many perturbing variables. The data nec-essary for the partition of causation are rarely available in field studies of forests subject to intermediate levels of air pollution. Our inability to accurately estimate pollution deposition and the complexity of ecosystem responses make it ex-tremely difficult or virtually impossible to estimate quantitatively air pollution stress effects on forest ecosystems distant from point sources or to link quan-titatively the emission of pollutants with regional air pollution damage to eco-systems.

Difficulty in predicting system behavior is not limited to ecosystem analysis but is common to most, if not all, fields dealing with complex systems like meteorology, seismology, public health, and economics. Ecological scientists asked to predict effects of regional air pollution on terrestrial ecosystems should follow the lead of macroeconomists, for example, who deal with systems beset with complexities almost as great as those encountered in natural systems. Faced with requests for advice on enormously important economic questions, econ-omists draw on circumstantial evidence and their world view of economics to render the best answer possible. Although there is often substantial disagreement among economists, attempts by individuals to render their best judgment are not inhibited by a lack of agreed-upon predictive macroeconomic models (Feld-stein et al. 1984).

Ecological scientists faced with requests for advice on important questions concerning the impact of regional air pollution should use the substantial base of circumstantial evidence that is available to them, for example the occurrence of air pollution symptoms in natural ecosystems, forest plantations, and crops; experimental evidence obtained from field and laboratory studies; levels of air pollutants reported from monitoring systems; data on accumulation of trace metals in soils; and basic knowledge about air pollution/organism interactions. This information along with the ecologists' knowledge of ecosystem structure, function, and development, and their world view of ecology should be used to render the best informed judgment possible at this time.

A world view shared by many ecologists projects that stress, including air pollution stress, will markedly increase during the next few decades. This increase will be in response to growth of world population and rising expectations and will take the form of increased demands for energy, natural resources, transportation, urban expansion, recreation, and agricultural, manufactured, and wood products (Holdgate et al. 1982). In terms of naturally occurring ecosystems these trends translate into an increase in a variety of stresses (personal communication, William R. Burch, Jr., Yale University, New Haven, CT, November 1984; and Barney 1974; Anonymous 1979; Hewett and Hamilton 1982, U.S. Forest Service 1981, 1982; Peterson 1984) in addition to air pollution stress. Substantial areas of natural ecosystems will be converted to agriculture, roads, urban development, mining sites, and water impoundments with a great loss of biotic regulation over energy flow and biogeochemical cycles. Further impacts on remaining forest land will come from increased harvesting pressure, more intensive harvesting methods, regeneration failures, and intensive forestry management using more pesticides and fertilizers. Still other impacts will result from rapidly increasing recreational activities that will result in more trails, campgrounds, and ski areas, an increasing population of off-road vehicles, heavier hunting and fishing pressure, and increased risk of fire. It is against this background that effects of current levels of air pollution stress and future air quality standards should be weighed. The health of ecosystems is determined by all the stresses acting on them!

To accommodate future stress, "ecosystems will need all the redundancy they can muster" to maintain even a modified equilibrium. Ecosystem redundancy should be reserved to buffer the stresses we can't control. In a few decades, we may have to accept a lowered species diversity and less biotic regulation of energy flow and biogeochemical cycles. Under these circumstances, it seems injudicious, at this time, to sacrifice ecosystem redundancy by weak air quality laws and programs that are designed to achieve short-term financial and economic objectives.

Whatever our world view, clearly it is necessary to improve our understanding of the air pollution problem with a continuing intensive research effort in ecology and related fields. However, the claim made by some scientists and bureaucrats that "more research is needed" before any ecological and landscape scientists' best informed judgment about the air quality–landscape interaction is rendered is not responsible. Policy makers need to act and should have the benefit of ecological and landscape scientists' best informed judgment. They should accord these judgments the consideration they deserve as coming from the investigators whose opinions are the best forecast of the future health of forest ecosystems.

3.8   Actions in Response to Air Pollution

In thinking about an immense problem such as regional air pollution and its impact on regional landscapes, the individual landscape scientist or manager may ask: "Is this problem in my domain?" And if it is, "What can I do about it?" To this author, it is in our domain and to ignore regional air pollution simply may be to preside over a vast scenario of creeping degradation, despite our best efforts to understand and manage local landscapes.

There are both direct actions in landscape planning and indirect actions on questions that affect landscapes that can be taken by scientists and managers to help ameliorate the regional air pollution problem.

The clearest direct action is through the conservation of energy. The conservation of energy is the best way to diminish air pollution. Greater efficiency in the use of energy in buildings, transportation, manufacturing, agriculture, and so forth is far more effective than emission control in controlling air pollution. Coal and oil that is not burned creates no pollution at all! Landscape scientists and managers can contribute to this by advocating, designing, and implementing low-energy landscapes. Low-energy landscapes are ones that run to the largest degree possible, commensurate with planning objectives, on solar energy. Optimum reliance on solar energy in landscape design not only contributes to the important goals of reducing the drain on global natural resources and lessening regional and local impacts of air and water pollution, but simultaneously provides the least expensive landscape management.

Indirect actions are those in which landscape scientists and managers become involved in the public debates over issues where the outcomes directly affect landscapes. There are many of these such as the current national debate over the strictness of air quality laws. Sometimes it is important to be aware of linkage in deciding whether or not to participate in a particular debate. For example, the recent decision of the Federal Government to lower mandated mileage requirements for General Motors and Ford was much more than an economic decision that promoted production of big cars (von Hippel 1985). It promoted greater use of steel and plastics, as well as greater use of gasoline. The decision resulted in greater operational cost to the American consumer, delayed development of more efficient cars, increased the drain on a finite natural resource, created more dependence on foreign oil, and, pertinent to this discussion, increased the production of air pollutants and greater stress on ecosystems.

Finally, landscape scientists and managers should pay attention to education and see to it that unifying concepts of ecology, ideas on landscape durability and sustainability, and environmental ethics are cornerstones in the training of new members of their profession. In the larger sense, they should be concerned with general education. Americans concerned with environmental quality profited enormously from the intense environmental debate of the 1960s, and early 1970s. Today middle-aged people who were subjected to that debate are supplying much of the leadership in our nation's effort to maintain environmental quality. Now the climate has changed! Young people are blitzed with well-

funded and effective advertising campaigns based on ideologic antienviron-
mentalism. A prevalent "me first" attitude and a steadily weakening public
expression of environmental concern coupled with a low-paying and shrinking
career potential in environmental work mean that fewer and fewer of our youth
are attracted to a life's work in environmentalism or even a strong interest in
the subject. Not only does this trend bode ill for the continued evolution of
enlightened landscape science, but the implications for all of us are extreme.

References

Andresen, A.M., Johnson, A.H., Siccama, T.G. 1980. Levels of lead, copper and zinc
in the forest floor in the northeastern United States. *J. Environ. Qual.* 9:293–296.
Anonymous 1979. Renewable natural resources: Some emerging issues. U.S. Government
Printing Office, Washington, D.C.
Anonymous 1982. Forest damage from air pollution. Publication series of the German
Federal Ministry of Nutrition, Agriculture and Forests. Series A, Applied Science,
Vol. 273. Agriculture Publishers Munster-Hiltrup.
Antonovics, J. 1968. Evolution in closely adjacent plant populations. V. Evolution of
self-fertility. *Heredity* 23:219–238.
Ayazloo, M., Bell, J.N.B. 1981. Studies on the tolerance to sulfur dioxide of grass
populations in polluted areas. I. Identification of tolerant populations. *New Phytol.*
88:203–222.
Barney, D.R. 1974. *The Last Stand.* Grossman Publishers, New York.
Bell, J.N.B., Clough, W.S. 1973. Depression of yield on rye grass exposed to sulphur
dioxide. *Nature* 241:47–49.
Benoit, L.F., Skelly, J.M., Moore, L.D., Dochinger, L.L.S. 1982. Radial growth reductions of *Pinus Strobus L.* correlated with foliar ozone sensitivity as an indicator
of ozone-induced losses in eastern forests. *Can. J. For. Res.* 12(3):673–678.
Bormann, F.H. 1976. An inseparable linkage: Conservation of natural ecosystems and
the conservation of fossil energy. *BioSci.* 26(12):754–760.
Bormann, F.H. 1982. Air pollution stress and energy policy, pp. 85–140. *In* C.H. Reidel
(ed.), *New England Prospects: Critical Choices in a Time of Change.* University
Press of New England, Hanover, NH.
Bormann, F.H. 1983. Factors confounding evaluation of air pollution stress on forests:
Pollution input and ecosystem complexity, pp. 147–166. *In* H. Ott and H. Stangl
(eds.), *Proceedings of the Karlsruhe Symposium on Acid Deposition: A challenge
for Europe.* Preliminary edition. XII/ENV/45/83. Commission of the European Com-
munities, Brussels, Belgium.
Bormann, F.H., Likens, G.E. 1979a. *Pattern and Process in a Forested Ecosystem.*
Springer-Verlag, New York.
Bormann, F.H., Likens, G.E. 1979b. Catastrophic disturbance and the steady state in
northern hardwood forests. *Am. Sci.* 67(6):660–669.
Bormann, F.H., Likens, G.E. 1985. Air and watershed management and the aquatic
ecosystem, chapter IX. *In* G.E. Likens (ed.), *An Ecosystem Approach to Aquatic
Ecology: Mirror Lake and its Environment.* Springer-Verlag, New York.
Bormann, F.H., Likens, G.E., Siccama, T.G., Pierce, R.S., Eaton, J.S. 1974. The export
of nutrients and recovery of stable conditions following deforestation at Hubbard
Brook. *Ecol. Monogr.* 44(3):255–277.
Bradshaw, A.D. 1976. Pollution and evolution, pp. 135–159. *In* T.A. Mansfield (ed.),
Effects of Air Pollutants on Plants. Cambridge University Press, New York.
Burgess, R.L. (ed.) 1984. *Effects of Acidic Deposition on Forest Ecosystems in the
Northeastern United States: An Evaluation of Current Evidence.* State University
of New York, College of Environmental Science and Forestry, Institute of Envi-
ronmental Program Affairs, Syracuse, NY.

Davis, D.D., Wilhour, R.G. 1976. *Susceptibility of Woody Plants to Sulfur Dioxide and Photochemical Oxidants: A Literature Review.* USEPA 600/3-76-102. Corvallis, OR.

Duchelle, S.F., Skelly, J.M., Sharich, T.L., Chevone, B.I., Yang, Y.S., Nellssen, J.E. 1983. Effects of ozone on the production of natural vegetation in a high meadow of the Shenandoah National Park of Virginia. *J. Environ. Mgmt.* 17:299–308.

Environmental Protection Agency. 1984a. Air Quality Criteria for Ozone and Other Photochemical Oxidants. Vol. III. EPA-600/8-84-020A, Corvallis, OR.

Environmental Protection Agency. 1984b. National Aerometric Data Bank. Research Triangle Park, NC.

Feldstein, M.S., Tobin, J., Meltzer, A.H., Ture, N.B. Looking ahead. *New York Times.* 1F,2F. Nov. 11, 1984.

Galloway, J.N., Likens, G.E., Hawley, M.E. 1984. Acid precipitation: Natural versus anthropogenic components. *Science* 226:829–831.

Galloway, J.N., Thornton, J.D., Norton, S.A., Volchok, H.L., McLean, R.A.N. 1982. Trace metals in atmospheric deposition: a review and assessment. *Atmos. Environ.* 16(7):1677–1700.

Gordon, A.G., Gorham, E. 1963. Ecological aspects of air pollution from an iron-sintering plant at Wawa, Ontario. *Can. J. Bot.* 41:1063–1078.

Grobstein, C. 1983. Should imperfect data be used to guide public policy? *Science '83* :18, December.

Heck, W.W., Taylor, O.C., Adams, R., Bingham, G., Miller, J., Preston, E., Weinstein, L. 1982. Assessment of crop loss from ozone. *J. Air Pol. Control Assoc.* 32(4):353–361.

Hewett, C.E., Hamilton, T.E. (eds.) 1982. *Forests in Demand: Conflicts and Solutions.* Auburn House, Boston, MA.

Hill, A.R. 1975. Ecosystem stability in relation to stresses caused by human activities. *Can. Geog.* 19(3):206–220.

Holdgate, M.W., Kassas, M., White, G.F. 1982. World environmental trends between 1972 and 1982. *Environ. Conserv.* 9(1):11–29.

Houston, D.B., Stairs, G.R. 1973. Genetic control of sulfur dioxide and ozone tolerance in eastern white pine. *Forest Sci.* 19:267–271.

Hutchinson, T.C. 1980. Impact of heavy metals on terrestrial and aquatic ecosystems, pp. 158–164. *In* P. Miller (coord.), *Proceedings of Symposium on Effects of Air Pollutants on Mediterranean and Temperate Forest Ecosystems.* Pacific SW Forest and Range Experiment Station. Gen. Tech. Report PSW 43, Berkeley, CA.

Johnson, A.H., Siccama, T.G. 1983. Acid deposition and forest decline. *Environ. Sci. Tech.* 17:294A–305A.

Jordan, M.J. 1975. Effects of zinc smelter; emissions and fire on a chestnut-oak woodland. *Ecology* 56:78–91.

Karnosky, D.F. 1977. Evidence for genetic control of response to sulfur dioxide and ozone in *Populus tremuloides. Can. J. For. Res.* 7:437–440.

Knabe, W. 1976. Effects of sulfur dioxide on terrestrial vegetation. *Ambio* 5:213–218.

Last, F.T. 1984. Direct effects of air pollutants singly and in mixtures on plants and plant assemblages, pp. 105–126. *In* H. Ott and H. Stangl (eds.), *Proceedings of the Karlsruhe Symposium on Acid Deposition: A challenge for Europe. (Preliminary Edition) XII/ENV/45/83.* Commission of the European Communities, Brussels, Belgium.

Likens, G.E. 1983. A priority for ecological research. *Bull. Ecol. Soc. Am.* 64:234–243.

Likens, G.E. 1984. Beyond the shoreline: A watershed-ecosystem approach. Verh. Internat. *Verein Limnol.* 22:1–22.

Likens, G.E., Bormann, F.H. 1974. Linkages between terrestrial and aquatic ecosystems. *BioSci.* 24(8):447–456.

Likens, G.E., Bormann, F.H., Pierce, R.S., Eaton, J.S., Johnson, N.M. 1977. *Biogeochemistry of a Forested Ecosystem.* Springer-Verlag, New York.

Likens, G.E., Butler, T.J. 1981. Recent acidification of precipitation in North America. *Atmos. Environ.* 15(7):1103–1109.

Ludwig, F.L., Shelar, Jr., E. 1980. Empirical relationships between observed ozone concentrations and geographical areas with concentrations likely to be above 120ppb. *J. Air Poll. Control Assoc.* 30:894–897.

Manning, W.J. 1979. Air quality and plants. *Science* 203:834.

Margalef, R. 1969. Diversity and stability: A practical proposal and a model of interdependence, pp. 25–38. *In Diversity and Stability in Ecological Systems*. Brookhaven Symposium in Biology No. 22.

Miller, P.R., Elderman, M.J., Kickert, R.N., Taylor, O.C., Arkley, R.J. 1976. *Photochemical Oxidants Air Pollutant Effects on a Mixed Conifer Forest Ecosystem: A Progress Report.* USEPA 600/3-77-104. Corvallis, OR.

National Academy of Science. 1977. *Ozone and Other Photochemical Oxidants*. Washington, D.C.

Neagele, J.A., Feder, W.A., Bryant, C.J. 1972. *Assessment of Air Pollution Damage to Vegetation in New England: July 1971–July 1972*. USEPA-R5-72-009. Waltham, MA.

Peterson, R.M. 1984. *Draft Environmental Impact Statement, 1985–2030*. Resource Planning Act Program. Forest Service, USDA.

Reifsnyder, W.E. 1984. Hydrologic process models. *In United Nations University Workshop on Forests, Climate and Hydrology—Regional Impacts*. St. John's College, Oxford, England.

Siccama, T.G., Bliss, M., Vogelmann, H.W. 1982. Decline of red spruce in Vermont. *Bull. Torrey Bot. Club* 109:162–168.

Skelly, J. 1980. Photochemical oxidant impact on Mediterranean and temperate forest ecosystems: Real and potential effects, pp. 38–50. *In P. Miller (coord.), Proceedings of Symposium on Effects of Air Pollution on Mediterranean and Temperate Forest Ecosystems: An International Symposium*. Pacific SW Forest and Range Expt. Sta. Gen. Tech. Report PSW-43.

Smith, W.H. 1981. *Air Pollution and Forests*. Springer-Verlag, New York.

Society of American Foresters. 1984. Acidic deposition and forests. SAF, Bethesda, MD.

Sprugel, D.G., Miller, J.E., Muller, R.N., Smith, H.J., Xerikos, P.B. 1980. Sulfur dioxide effects on yield and seed quality in field-grown soy beans. *Phytopathology* 70:1129–1133.

U.S. Congress, House, Committee on Energy and Commerce. 1982. *Acid Precipitation*, part 2, pp. 98–99, 134–136. Hearings, Oct. 20, 1981, 97th Congress, 1st Session, Gov't Printing Office, Washington, D.C.

U.S. Forest Service. 1981. *An Assessment of the Forest and Range Land Situation in the United States*. Forest Resource Report No. 22, Department of Agriculture. U.S. Government Printing Office, Washington, D.C.

U.S. Forest Service. 1982. *An Analysis of the Timber Situation in the United States, 1952–2030*. Forest Resource Report No. 23, Department of Agriculture. U.S. Government Printing Office, Washington, D.C.

Ulrich, B., Pankrath, J. 1983. *Effects of Accumulation of Air Pollutants in Forest Ecosystems* Reidel, Boston, MA.

Usher, R.W., Williams, W.T. 1980. Assessment of pollution injury to eastern white pine in Indiana. *In O. Loucks, R. Armentano, R.W. Usher, W.T. Williams, R.W. Miller, and L.T.K. Wong (eds.), Crop and Forest Losses Due to Current and Projected Emissions from Coal-Fired Power Plants in the Ohio River Valley* (ORBES). USEPA R-805588.

von Hippel, F. 1985. Automobile fuel efficiency (letter). *Science* 228:263–264.

Wang, D., Karnosky, D.F., Bormann, F.H. 1986. Effects of ambient ozone on the productivity of *Populus tremuloides* Michx. grown under field conditions. *Can. J. For. Res.* 16:47–55.

Weathers, K.C., Likens, G.E., Bormann, F.H., Eaton, J.F., Bowden, W.B., Andresen, J.L., Cass, D.A., Galloway, J.N., Keene, W.C., Kimball, K.D., Huth, T., Smiley,

D. 1986. A regional acidic cloud/fog water event in the eastern United States. *Nature* 319:657–658.

Williams, W.T. 1980. Air pollution disease in the California forests: A base line for smog disease on ponderosa and Jeffrey pines in the Sequoia and Los Padres National Forests. *California Environ. Sci. Technol.* 14:179–182.

Williams, W.T., Williams, J.A. 1984. Smog decline of ponderosa pine forests worsens in southern Sierra Nevada. Abstract Second Biennial Conference of Research in California's National Parks. University of California, Davis.

Williams, W.T., Wong, L. 1980. A review of mechanisms of action of ozone and sulfur dioxide pollutants on crops and forests. *In* O. Loucks, T. Armentano, T.W. Usher, W.T. Williams, R.W. Miller, and L.T.K. Wong (eds.), *Crop and Forest Losses due to Current and Projected Emissions from Coal-Fired Power Plants in the Ohio River Valley.* (ORBES). USEPA R-805588.

Woodwell, G.M. 1970. Effects of pollution on the structure and physiology of ecosystems. *Science* 168:429–433.

Woodwell, G.M., Whittaker, R.H., Reiners, W.A., Likens, G.E., Delwiche, C.C., Botkin, D.B. 1978. The biota and the world carbon budget. *Science* 199:141–146.

Zhao, D., Xiong, J., Wu, Y. 1985. Acid rain in China: Its formation and effects. Presented at the U.S.–China Air Pollution Ecological Effects Workshop. November 1–4, 1985. Jiangsu Institute of Botany, Nanjing, China.

4. Parasites, Lightning, and the Vegetation Mosaic in Wilderness Landscapes

Dennis H. Knight

4.1 Introduction

The vegetation mosaic in any landscape is a function of environmental variation and historic disturbances, whether caused by humans or other factors. Many studies have focused on species composition in relation to environmental gradients, and secondary succession is one of the oldest themes in ecology (Fig. 4.1). Usually the study areas selected for research of this nature have been dispersed in agrourban landscapes and have included woodlots, small tracts of prairie, or other relatively small, homogeneous stands of vegetation. Although the ecological significance of the spatial positioning of the communities or patches in the agrourban matrix has been largely ignored, there seems to be a growing interest in determining how the number, variety, and juxtaposition of the patches influence the frequency and spread of disturbance as well as other landscape characteristics, including wildlife habitat, biotic diversity, nutrient retention, and productivity (Pickett and White 1985; Forman and Godron 1986).

Ecological research on landscapes, with its focus on the causes and significance of patterns beyond the scale of single stands or isolated communities, has been stimulated in part by human-caused landscape fragmentation. Over much of the earth's surface the natural mosaic is no longer visible, with "islands" of woodland or prairie interspersed in a "sea" of agriculture (Pickett and Thompson 1978; Ewel 1980). Elsewhere the natural mosaic is still apparent, but fragmentation is occurring rapidly due to timber harvesting (Harris 1984),

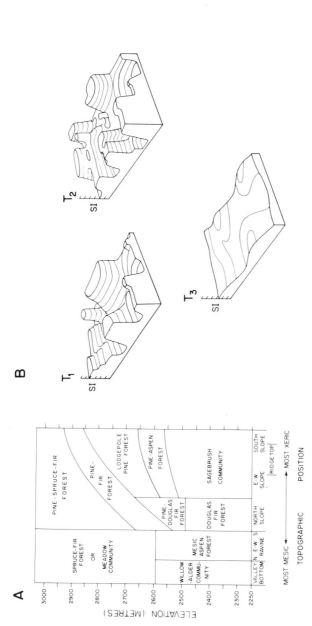

Figure 4.1. An approach for illustrating vegetation distribution (A) and vegetation dynamics (B) in landscapes. The diagram on the left, from Romme and Knight (1981), is an abstraction that shows where different vegetation types most commonly occur with regard to topographic position and elevation, but such diagrams do not portray the actual vegetation mosaic as it would appear on an aerial photograph. (B) After Pickett (1976), illustrates the shifting mosaic concept, with T₁ portraying the vegetation mosaic at one time, T₂ at some later time, and T₃ after a major disturbance. The vertical axis (SI) is some measure of successional development, and the other two axes are longitude and latitude. Although it illustrates very nicely the shifting mosaic concept, the diagram does not incorporate the environmental variability that always exists across landscapes. A major challenge for landscape ecologists working on wildlands is understanding the simultaneous effects of patch dynamics and environmental gradients on the nature of the vegetation mosaic. Reproduced with permission of the Ecological Society of America (A) and the University of Chicago Press (B).

for example in the Amazon Basin, the Pacific Northwest, and the Rocky Mountain Region (Fig. 4.2). Questions have arisen about the significance of preserving woodlots or hedgerows adjacent to croplands, the optimal size of woodlands or clearcuts, and the economic impacts, both in the near future and in the long term, of preserving a patchy landscape compared with one that is more homogeneous. Discussions focusing on these questions have ranged from early attempts to evaluate the significance of "edge" for wildlife (Leopold 1933) to more recent attempts to apply the theory of island biogeography to terrestrial landscapes (Burgess and Sharpe 1981; Harris 1984). Humankind has a significant effect on landscape pattern. In the interest of long-term natural resource management, what is the significance of the patterns that are being created? How do natural landscape mosaics compare with those created by humans, and what ecological information should landscape architects, farmers, and forest managers consider as they continue their work?

Only a few questions of this nature can be addressed in this chapter. My objective is to explore the potential effects of certain insects, fungi, and parasitic angiosperms on landscape pattern in forested wildlands, with the belief that studying such interactions can lead to the improved management of landscapes

Figure 4.2. As with the original forested wildlands in other parts of the world, the forests in the western and northern parts of North America are being fragmented by timber harvesting. New landscape mosaics are being created. Whereas timber harvesting is now done more frequently in ways that minimize visual impact, is there an ecological rationale for prescribing one vegetation mosaic over another? Such questions represent a major challenge for landscape ecologists.

in which forests are being fragmented by timber harvesting or homogenized due to fire suppression. I'll focus on parasites (*sensu lato*) and fire because (1) their interaction is interesting and widespread, (2) both can have significant effects on vegetation mosaics, and (3) both are often subjected to control measures. I emphasize the mosaic in wildlands because wilderness landscapes were the setting for the evolution of our biota and are changing due to human influences. Understanding the causes and significance of the vegetation mosaic in wildlands may help in the design of domesticated landscapes. Coniferous forests are emphasized because of the literature available, but brief comparisons will be made to other vegetation types.

4.1.1 Nature of the Shifting Mosaic

Studying wilderness landscapes presents a variety of challenges and opportunities. Unlike agrourban settings, the boundaries between the patches of the mosaic often are less regular and less contrasting (Forman and Godron 1981, 1986). Consequently, study area selection can be frustrating because the whole landscape presents so many possibilities. The actual mosaic over square kilometers or whole townships becomes interesting in itself, and the first impulse is to draw a map. Furthermore, the dynamics of the mosaic become as interesting as the dynamics of the patches (Romme and Knight 1982). Often it is not at all clear whether the border between two patches is attributable to environmental factors or disturbances at some time in the past, and indeed, understanding the interplay of environment and history is one of the primary challenges for landscape ecologists (Pickett and White 1985). It is tempting to characterize the shifting mosaic of wilderness landscapes as being kaleidoscopic in nature, with the potential of each patch type occurring anywhere on the landscape at some time in the future, but such areas are the exception. Many factors are involved (Table 4.1) and some parts of the mosaic may change hardly at all during long periods of time (Fig. 4.3).

Fire is a very important exogenous disturbance that affects the vegetation mosaic in many areas (Wright 1974; Heinselman 1981a, 1981b; Minnich 1983; Habeck 1985; and others), and its spread is affected by a variety of factors. For example, the species composition and fuel complex may not foster the spread of a fire once started (Minnich 1983; Despain 1985). If such patches are intermingled with patches that are more flammable, the size of any particular burn may be limited. Furthermore, the site of ignition must be considered. If ignition occurs high on a slope, or just upwind of an aquatic habitat, the extent of burning may be less compared with when ignition occurs lower or some distance from natural fire breaks. The mosaic may also be affected by shifting winds, infestations of parasites that affect flammability, the degree of topographic dissection, and climatic conditions that extinguish the fire, thereby limiting its spread. Of course, factors other than fire may affect the mosaic as well (Sprugel 1976; Reiners and Lang 1979).

Fires on any particular part of the landscape are stochastic phenomena and, to some extent, are biotically controlled, (i.e., time-dependent, Heinselman

Table 4.1. Some Factors Affecting the Spread of Fire and the Resulting Vegetation Mosaic in Wilderness Landscapes

Vegetation mosaics
  Patch diversity (number of different kinds of patches)
  Patch size structure (number of patches by patch size class)
  Patch juxtaposition (proximity of one patch type to another)
  Patch structure (fuel complex, species composition)
Biotic influences on flammability (positive or negative)
  Drying
  Nature of the fuel complex
Topography
  Slope position of ignition (less area may burn if ignition occurs on upper slopes)
  Relief
  Degree of dissection
Barriers to fire spread
  Aquatic habitats
  Ridges, valleys
  Patches of vegetation with low flammability
Climatic factors
  Rainfall (reduces probability of fire starts or may extinguish a fire)
  Drought
  Variable wind direction and intensity
Ignition source and frequency
  Lightning
  Fire brands
  Anthropogenic

1981a, 1981b). One objective of this chapter is to consider the extent to which ignition by lightning and subsequent fire spread are affected by organisms, or, to phrase the objective in the context of the symposium theme, to determine the role of landscape heterogeneity caused by parasites (a biotic factor) in the spread of disturbance caused by lightning (a physical factor). Some of the literature that I review suggests that epidemic outbreaks of certain organisms may either increase or decrease the flammability of coniferous forests and that lightning itself may affect the pattern of parasite infestations.

As with single-stand studies, ecologists have sometimes suggested that a steady state or equilibrium condition could develop for landscapes, meaning that the proportion of the landscape in various vegetation types (or patch types) remains about the same even though the location of each type changes with time. Bormann and Likens (1981) proffered the shifting mosaic steady-state hypothesis for northeastern deciduous forest landscapes, where large catastrophic fires occur very infrequently. On the other hand, Romme (1982) found a nonequilibrium mosaic on 73 square kilometers in Yellowstone National Park characterized by large fires; studies are currently underway to determine if larger portions of Yellowstone could be viewed as having a shifting mosaic steady state (W.H. Romme and D.G. Despain, personal communication). Hemstrom and Franklin (1982) studied 530 square kilometers of forest in Mount

Figure 4.3. A drawing based on an aerial photograph taken in Yellowstone National Park that illustrates how fire and environmental factors combine in determining the vegetation mosaic in wilderness landscapes. The meadows, shown as stippled areas, are relatively stable in their location, occurring where edaphic conditions are less favorable for tree growth. The shape of the burned areas (white) can be attributed to a variety of factors including fuel discontinuities, the location of the lightning strike, fire brands (causing the smaller burned patches), shifting winds, and other factors such as those listed in Table 4.1. The fire burned until extinguished in the fall by rain and snow.

Rainier National Park and found no indication of an equilibrium landscape. Botkin (1980) found no a priori reason for equilibrium landscapes to occur, and paleoecological studies (Webb 1981; Davis 1981) suggest that the time scale of succession may overlap the paleoclimatological time scale more than ecologists are accustomed to admitting. Thus, the equilibrium concept may be of little value in many areas, whether for communities or landscapes.

4.1.2 A Brief Overview of Past Research on Wildland Fire

Of all the disturbances that affect the nature of vegetation mosaics, fire is undoubtedly the best studied. Many books have focused on the ecological effects of fire (e.g., Kozlowski and Ahlgren 1974; Mooney et al. 1981; Wein and MacLean 1983; Chandler et al. 1983; Pyne 1984; Lotan et al. 1985a), and various techniques have been devised for determining the mean return interval (Arno and Sneck 1977; Arno 1980; Swain 1980; Minnich 1983) and rate of spread

(Rothermel 1983; Albini 1983, 1984) of fires in different regions. With regard to North American wildland fires, the reviews of Wright (1974), Kilgore (1981), and Heinselman (1981a, 1981b, 1985) are particularly noteworthy.

With few exceptions, ecological studies on fire have focused on (1) postfire successional patterns, (2) fire frequency in relation to habitat and climate, (3) plant adaptations for surviving fire or taking advantage of conditions created by fires, (4) the significance of fire suppression, (5) factors affecting the spread of fire, (6) the effects of fire on ecosystem processes, and (7) the use of fire as a management tool. All have relevance to the objectives of landscape ecology and many emphasize the role of fire in perpetuating a mosaic of community types and age classes (e.g., Rowe 1979; Forman and Boerner 1981; Heinselman 1981a, 1981b; Minnich 1983; Habeck 1985). Although most studies have been done on specific stands with relatively little attention to the significance of the mosaic or combinations of perturbations contributing to fire susceptibility, the development of regional landscape models (Johnson 1977; Kessell 1979; Romme 1982; Weinstein and Shugart 1983; Shugart and Seagle 1985), combined with significant advances in understanding flammability (Brown 1975; van Wagner 1977; Rothermel 1983; Albini 1984; van Wagtendonk 1985; Andrews 1986), have set the stage for research on fire ecology from a landscape perspective.

The gradient modeling approach of Kessell (1976a, 1976b, 1979) and his associates was one of the first landscape studies with a focus on fire. Kessell found relationships between fuel characteristics and various environmental gradients in Glacier National Park, and then conducted a detailed hectare-by-hectare inventory of portions of the park, measuring the critical environmental factors in each landscape "cell." With this information accessible by computer, park managers could then predict the probability that a fire starting in one cell would spread to adjacent cells. Kessell recognized that predicting weather conditions and the potential for crowning and spotting by fire brands were problematical, but his approach illustrated a new application of traditional methodology to questions about landscapes.

Other landscape models also require dividing the terrain into cells. For example, Hett (1971), Romme (1982), and Weinstein and Shugart (1983) developed models that permitted the calculation of changes in landscape characteristics over time, for example, the percentage of the area in various successional stages. Although they provided insights into landscape dynamics, these studies did not establish the importance of patch proximity (or juxtaposition), and, as with the studies of Kessell, they did not incorporate the causes of flammability. As described below, there are a variety of processes that affect flammability and are sufficiently well understood to include in future, less empirical models that portray the dynamics of the vegetation mosaic.

4.2 The Natural History of Flammability

Flammability can be defined as the relative ease with which a substance ignites and sustains combustion (Wein and MacLean 1983, p. 301). Although flam-

mability would appear to be a simple function of fuel accumulation during
succession, there are various factors that must be considered to gain a more
mechanistic understanding of why some areas burn more frequently or more
intensely than others. This is true regardless of whether the focus is on factors
affecting the spread of fire within one patch of the mosaic or from one patch
to another. Great strides have been made in measuring flammability, especially
fuel loadings (Brown et al. 1982), and various models exist for predicting the
rate of fire spread (Rothermel 1983; Albini 1984; Andrews 1986). In this section
I review some of the possible mechanisms whereby flammability increases or
decreases (Fig. 4.4), particularly with regard to parasites in forest ecosystems.

Flammability is a function of drying, wind, plant species composition, the
fuel complex where ignition occurs, and the fuel complex of adjacent patches
of the landscape mosaic (Fig. 4.4). Each of these factors is affected in turn by
others; for example, the fuel complex can become more flammable through the
effects of suppression mortality, plant growth, parasites, or wind storms (Lotan
et al. 1985b). Such factors can modify fuel continuity, porosity, surface/volume
ratio, and dead/live biomass ratio—all of which affect flammability. One fire
may reduce the probability of another (Lotan et al. 1985b), as in the case where

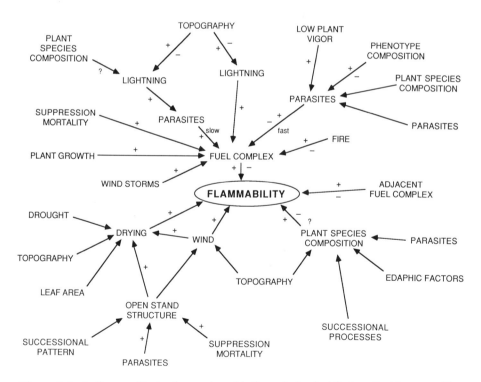

Figure 4.4. A diagram illustrating various biotic and physical factors that have positive
(+) or negative (−) effects on flammability, as discussed in the text. Some factors are
included more than once because they may contribute to the development of flammability
in various ways.

forest fuel continuity does not extend into the canopy because of frequent surface fires.

4.2.1 Drying

Climatic drought, microclimatic conditions caused by topography, rapid transpiration due to the development of high leaf area, a more open stand structure,
or the proximity of open areas can lead to drying that greatly increases flammability if all other conditions are favorable as well (Fig. 4.4). Open stand structure
can be caused by suppression mortality or wind storms, but Furyaev et al.
(1983) suggested that the spruce budworm can be another important cause in
the boreal forest of eastern Canada. They observed that portions of the spruce-
fir forest often remain quite moist throughout the year, and that flammability
could increase after budworm infestations that open up the forest canopy, causing more drying of understory fuels due to more air movement and light pen-
etration. They described a similar phenomenon caused by other insects for
Siberian fir forests in the Soviet Union, and Brown (1975) suggested the potential
for drying in the northern Rocky Mountains after mountain pine beetle infestations. Furyaev et al. (1983) lamented the lack of data to test their Insect-Fire
Hypothesis, and acknowledged that changes in the fuel complex after infestations could be important as well. However, even if the forests would burn
without the insect infestations (e.g., during infrequent drought years), the insects
probably do affect the mean fire return interval as well as fire intensity, the
size and shape of the area burned, and postfire successional patterns.

4.2.2 The Fuel Complex

Many studies have identified insects and other pathogens as important factors
in increasing fuel continuity, porosity, dead/live ratios, and surface/volume ratios
(Wright and Heinselman 1973; Brown 1975; Johnson and Denton 1975; Kilgore
1981; Schowalter et al. 1981, 1985; Heinselman 1983), but recent research has
provided more details on the mechanisms involved, some of which are coun-
terintuitive. For example, outbreaks of mountain pine beetle in western North
America may create patches of dead trees which appear very flammable. However, a recent study (Romme, Knight, and Fedders, in preparation) found that
whereas flammability may increase for a year or two while the dead leaves
persist on the trees, leaf fall is quite rapid and fuel continuity in the canopy
may actually decrease for 20 years or more after the infestation. The accelerated
growth of understory trees (Romme et al. 1986) and the toppling of the insect-
killed trees may eventually enhance flammability, but not in the initial decade
as is often thought.

Other factors must be considered as well in evaluating the effect of insects
on the fuel complex. Stocks (1985) studied the effect of spruce budworm outbreaks on flammability in Ontario, observing that fire spread may be restricted
for a few years by the moist understory vegetation which sometimes proliferates

after canopy defoliation. Another factor to consider in some areas may be reduced transpiration due to lower leaf area, which could preserve higher fuel moisture levels as a result of lowered rates of soil drying. Prescribed burns on the humid southeastern coastal plain have been observed to bypass patches of forest killed by the southern pine beetle, perhaps because of higher moisture content or lower concentrations of the flammable extractives in older fine fuels (T.M. Williams, personal communication).

Other insects may have similar effects, whether in terms of enhancing flammability or delaying its development, and, of course, fire suppression has been suggested as favoring conditions that lead to insect outbreaks, which in turn may enhance the probability of the next fire (Amman 1977; Peterman 1978; Schowalter et al. 1981, 1985; McCune 1983; Heinselman 1983; Waring and Schlesinger 1985; Stocks 1985). Although weather conditions favorable for hot fires may override the more subtle effects of insects, the interaction between insect outbreaks and fuel dynamics (flammability) appears to be significant. More research is needed to quantify the relationships.

Mistletoe, a common parasitic angiosperm on conifers, also affects the fuel complex, especially in terms of fuel continuity. Abnormal branching on infected trees creates living clumps of flammable fine fuels that, furthermore, serve to catch and accumulate litterfall (Hawksworth 1975; Alexander and Hawksworth 1975; Zimmerman and Laven 1984). Flammable resin-filled burls may also form (Wicker and Leaphart 1976). A fireball often results when such branches are ignited, which in turn may ignite the canopy, and flaming "witches brooms" may roll downhill or be carried aloft in convection columns, thereby causing spot fires (Alexander and Hawksworth 1975; Albini 1983). Different fuel configurations may be caused by mistletoe depending on whether the brooms are retained on the tree, as in ponderosa pine, or if they fall to the ground as in western larch and Douglas fir (Wicker and Leaphart 1976).

As with insects, fire suppression may enhance the spread of mistletoe (Wicker and Leaphart 1976; Zimmerman and Laven 1984), thereby hastening the development of conditions favorable for a larger, more intense fire. Thus far there is no indication that mistletoe increases or decreases tree susceptibility to other diseases or insects, for example, the mountain pine beetle (Hawksworth et al. 1983), but reduced tree vigor from any cause may increase susceptibility to parasitism (Berryman 1972; Raffa and Berryman 1982; Cates et al. 1983; Waring and Pitman 1983).

Trees damaged by lightning, wind, or fire may serve as epicenters for the spread of parasitism in coniferous forests, which may slowly enhance flammability (Taylor 1973; Schmid and Hinds 1974; Schowalter et al. 1981). One particularly appropriate example of such interactions was reported by Coulson et al. (1983) for the southern pine beetle. Lightning-struck trees may not ignite, and often they are not killed, but the beetles are able to locate damaged trees, probably because of increased terpene emissions (Krawielitzki et al. 1983). While ice storms and wind could also be important in creating epicenters, Coulson et al. think that lightning is the only disturbance that is sufficiently frequent to be a primary component in the evolution of the beetle's life history. In some parts of the beetle's range, lightning strikes may occur at the rate of 45 to 180 per square mile per year (Coulson et al. 1983).

Other studies have reported the mortality of groups of trees as a result of nonpyrogenic lightning, with the cause sometimes thought to be root rots (Taylor 1973). In addition to being epicenters for the spread of parasitism, whether by insects or fungi, wind damage may increase as gaps are created in the canopy. Flammability may increase in either case. Interestingly, Royama (1984) found no evidence for the importance of epicenters during outbreaks of the spruce budworm.

Fire itself may create favorable conditions for the perpetuation of some tree parasites, primarily because of the fire-damaged survivors that always occur. Geiszler et al. (1980) and Gara et al. (1985) concluded that fire-damaged lod-gepole pine are more prone to certain fungal infections, and that these trees are more susceptible to mountain pine beetle infestation. Schowalter et al. (1981) hypothesized a similar situation with the southern pine beetle in the coniferous forests of the southeast. The beetle population reaches epidemic numbers when stand conditions are right, leading to rapid changes in the fuel complex and the eventual enhancement of flammability. The next fire creates more fire-damaged survivors, which serve as epicenters for the next wave of fungal and beetle parasitism (Fig. 4.5). Other studies also suggest how pathogens may predispose trees to beetle infestations, for example, root rots (Gara et al. 1985; Tkacz and Schmitz 1986). Various mechanisms may be involved including (1) the fungal oxidation of alpha-pinene to trans-verbenol, a beetle aggregating pheromone; or (2) the reduced vigor of infected trees, which may limit their capacity to resist insect invasion.

Fungi may also play a more direct role in the development of flammability, for example, root rot (Dickman 1984), blister rusts, and oak wilt fungi. Some depend on insects as vectors, suggesting a mutualistic relationship (for example, the case of blue stain fungi and the mountain pine beetle; Amman 1978).

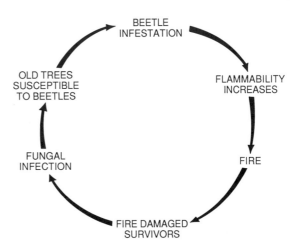

Figure 4.5. An illustration of the fungus/bark beetle/fire interaction proposed by Geiszler et al. (1980) and Gara et al. (1985).

4.2.3 Plant Species Composition

A primary objective of many ecological studies has been to determine the factors affecting community species composition, and some of these studies are relevant to the natural history of flammability. In addition to illustrating further the effect of parasites, this research has provided evidence that certain species are more flammable than others. Thus, cycles of flammability may depend on the species composition of the live biomass as well as drying conditions, wind, and characteristics of the detritus (Fig. 4.4).

With regard to parasites, some tree species are more susceptible to infection than others. For example, the pines are much more susceptible to mistletoe than are the various species of spruce and fir, and thus mistletoe will play a role in causing flammability only when pines are common. Because the pines are usually seral species, Tinnin (1984) suggested that mistletoe tends to perpetuate conditions favorable for the persistence of pine, even to the extent of increasing the probability of the next fire, which perpetuates an abundance of the pines on which the parasite depends. Fire suppression seems to have led to an increased abundance of mistletoe (Alexander and Hawksworth 1975; Wicker and Leaphart 1976; Zimmerman and Laven 1984), and probably a more rapid development of flammability. It appears as though the spread of mistletoe serves as a feedback mechanism for perpetuating the species on which it depends. Of course, spruce-fir forests are often quite flammable, even without parasites, due to the higher fuel continuity which exists in all-aged or uneven-aged stands.

Another example of differential susceptibility is the case of the spruce budworm. Balsam fir is more susceptible than white spruce, which in turn is more susceptible than black spruce. Jack pine, aspen, and birch are not susceptible. Thus, conditions favorable to the establishment of the nonsusceptible species could delay the occurrence of fire unless other mechanisms exist for accelerating flammability over and above what would be expected by biomass accumulation without perturbations. The western spruce budworm may have similar effects (J.K. Brown, personal communication), and Dickman (1984) describes a comparable example for the root rot *Phellinus weirii*, which infects western hemlock more readily than lodgepole pine in Oregon. With sufficient time between fires, the more susceptible hemlock may replace the pine during succession, with the root rot then accelerating the development of flammability.

Emphasis on species differences should not overshadow differences in genotypic and phenotypic variability within a species. Various investigators present evidence that some individuals of a species may be more susceptible to parasitism (Edmunds and Alstad 1978; McDonald 1981; Cates et al. 1983; Dinoor and Eshed 1984), and several studies have found that environmental conditions leading to reduced vigor will increase susceptibility (Raffa and Berryman 1982; Cates et al. 1983; Matson and Waring 1984; Waring and Pitman 1983, 1985). Such studies have great relevance to forest management programs in which pest control is desirable.

Another dimension to the effect of species composition on flammability was outlined with the hypothesis that some species may evolve structural or chemical characteristics that increase flammability (Mutch 1970). While ecologists usually focus on adaptations for maximizing photosynthesis or minimizing the effects of herbivory or fire, Rundel (1981) suggests that fire-dependent plants could also have evolved features that increase fire frequency. Such features include the production of more flammable tissues, the development of live tissues with a relatively low moisture content, or fuels that decay more slowly and thereby lead to a more rapid fuel buildup. Testing such hypotheses is difficult because what is perceived as an adaptation for enhancing flammability may actually be an adaptation for some other environmental stress (Christiansen 1985).

4.2.4 The Adjacent Fuel Complex

The susceptibility of any particular patch in the landscape mosaic to being burned is, to some degree, determined by the flammability of the adjacent fuel complex. Many have noted how flammability tends to increase with forest age, but Heinselman (1985) observed that young stands may burn as well once a major fire is underway, and Weinstein and Shugart (1983) made the assumption that when one unit of a landscape mosaic becomes highly flammable, adjacent units become more susceptible to "fire invasion." Although this is surely true to some degree, Despain and Sellers (1977) observed that forest fires in Yellowstone National Park became much less intense, or were even extinguished, when the advancing flames reached a tract of young forest with a much less flammable fuel complex. Minnich (1983) noted the same phenomenon for chaparral in California. Such observations suggest that the juxtaposition of young and old stands could reduce significantly the flammability of the landscape as a whole, a proposition commonly cited when discussing the landscape homogenization that could result from fire suppression (Minnich 1983; Habeck 1985). Heinselman (1985) suggested that the probable path of fires often can be predicted by mapping stands of old growth downwind from ignition points, which brings to mind the gradient modeling approach of Kessell (1979). Because such stands often are not large, fire spread may be restricted by stand age and/or less flammable patches of the vegetation mosaic (Despain and Sellers 1977; Romme and Knight 1982; Minnich 1983; Despain 1985). Some forest types may, in fact, function as natural fire breaks, for example, stands of red alder in the Pacific Northwest (J.F. Franklin, personal communication).

More is involved, however, than simple contrasts in forest age, and in some areas age may not be that important (Brown 1975; Lotan et al. 1985b). Heinselman (1985) and van Wagner (1983, 1985) suggested that young boreal forests are as flammable as older forests, and Franklin and Hemstrom (1981) noted that young forests (25 to 75 years) in the Pacific Northwest appear to be more flammable than older forests. Fuel continuity may be higher in younger forests, when more of the fuel is near the ground and, consequently, less heat is required to ignite the canopy. Van Wagner (1978) noted that the rate of fire spread could be highest in young stands, lowest in mature stands, and high again in very old stands. Thus, in boreal forests and elsewhere the proximity of young stands could greatly facilitate fire spread into old stands, or vice versa.

Fire spread to other landscape patches also can occur through spotting, that is, the wafting of "fire brands" by convective columns to some point downwind (Albini 1983; Stocks 1985; Pyne 1984). Although common, stand ignition in this manner depends more on the intensity and duration of "fire brand showers," and on in situ flammability, than on the adjacent fuel complex.

Understanding fire spread from one patch to another is central to the theme of landscape ecology, regardless of whether the goal is characterizing the flammability of the entire landscape or one patch in the mosaic. The variety of patch types and patch juxtaposition seem to be as important as the factors discussed previously which affect in situ flammability. The extent to which fire spreads from one patch to the next will depend on the contrast in fuel conditions between the two patches, with fire spread occurring until the heat generated is inadequate for the pyrolysis and ignition of additional fuel particles (as discussed below). Of course, spotting by fire brands may negate the effect of fuel discontinuities if the brands can be wafted to a more flammable patch downwind.

4.3   The Natural History of Ignition

Ignition of the fuel complex becomes increasingly probable as the level of flammability increases, whether due to stand age or other physical and biotic factors, and at some point a fire is initiated by lightning or humankind. Although in some areas the frequency of fire may be affected by human activities, Heinselman (1973) and van Wagner (1983, 1985) suggested that the ignition source is relatively unimportant compared with the fuel loadings and weather at the time. Others have suggested that the influence of native North Americans has been underestimated (e.g., Gruell 1985), arguing that the less extensive coverage of closed forests in presettlement times was due to more frequent anthropogenic fires. Such fires could have been caused by lightning as well, even in grasslands (Komarek 1964), but regardless of the ignition source their effect probably was to create more frequent surface fires which, in some areas, may have slowed the development of flammability. The presence of humans, and their rationale for starting fires, surely represents another biotic influence on the shifting landscape mosaic. But can the same be said for ignition by lightning? Do the biota facilitate ignition, over and above their effects on the development of flammability, and are lightning strikes affected by the biota? Information pertaining to these questions is reviewed in the following sections.

4.3.1   Ignition

Before combustion can occur, sufficient heat is necessary to dehydrate the fuels; initiate the pyrolysis (volatilization) of various organics, including terpenes, fats, oils, and waxes; and ignite the gases produced by pyrolysis, thereby creating flaming combustion (Rundel 1981; Pyne 1984). Once the fuels have been dehydrated, ignition usually occurs at 500 to 700°C. Glowing combustion or smoldering may occur as well as flaming combustion, and is especially important because it produces more heat per gram of fuel, causes more root damage, accounts for most forest floor combustion, and provides a persistent source of ignition, for example, by sustaining the combustion of fire brands or smoldering logs (Taylor 1973; Agee 1981; Minnich 1983; Gara et al. 1984, 1986) which may be the ignition source for large fires after long delays. Fire spread will depend, of course, on the degree of flammability that has developed at the time, and will continue until fuel characteristics, oxygen, or heat fall below critical levels. The fact that these conditions are not easily maintained accounts for the observation that most forest fires burn less than a hectare (Pyne 1984).

There are various ways in which biotic influences or biotic characteristics could influence the amount of heat required for ignition. To illustrate, some plant materials do not dry as rapidly as others, due perhaps to better stomatal control or other mechanisms for maintaining water within the plant tissues (which may be a function of habitat). For example, van Wagner (1977) suggested that crown fires may not develop in aspen forest as readily as in coniferous forests because plant moisture content usually remains above 140%, compared with the 70 to 130% moisture which supports crown fires in adjacent conifers. Thus, the heat required for aspen dehydration could exceed the heat generated by the ignition source. Dead forest fuels are often classified into 1-, 10-, 100-, and 1000-hour categories, a classification based on the estimated amount of time required for dehydration to a certain level when subjected to drying (Rothermel 1983; Pyne 1984).

Also, different plant species or different plant materials could produce more or less flammable gases per British thermal unit (btu). For example, decaying wood gives off gases more readily than the wood of intact trees, even if moist, and some species apparently produce more volatile organics than others (Rundel 1981). It seems plausible that parasites could facilitate the production of flammable gases.

Most ignitions in forests seem to occur first in the forest floor or litter layer, probably because the fine fuels there are more compact and continuous. Van Wagner (1977) defines the conditions under which three different kinds of crown fires can develop, noting that usually a surface fire is necessary first because the crown by itself does not generate enough heat for dehydration, pyrolysis, and combustion. Wind is important for the starting and maintenance of crown fires, as is the height above ground to a minimum fuel bulk density (the fuel ladder concept). Thus, as discussed in the context of flammability, various biotic factors can have a significant effect on whether or not a crown fire develops.

4.3.2 Lightning

Bolts of lightning strike the earth thousands of times every day (182 million strikes per year for the earth), but less than 1% start fires (Taylor 1973). In part this low percentage can be attributed to conditions of low flammability at the place and time of the strike, in particular the fact that many strikes occur during rain storms, but the nature of the lightning strike is important also. All strikes have tremendous amounts of energy, but some, the long-continuing strikes which last about 100 msec, are hotter than "fast lightning" and are of sufficient duration to increase dramatically the chances of dehydration, pyrolysis, and flaming. Such strikes have sufficient energy to rupture a live conifer, creating a "shower" of fine fuel particles which undergo pyrolysis in msec and flash into a fireball that in turn ignites other fine fuels (Taylor 1973). In western North America, where up to 70% of the fires are caused by lightning, less than 20% of the lightning strikes are of the long-continuing variety (Taylor 1973). Of course, as discussed previously, all lightning strikes may have an indirect effect on flammability by creating additional fuel or providing a favorable environment for the rapid population increase of a parasite. By various means, lightning and/or parasites alter ecosystem structure in ways that affect flammability.

But is it reasonable to think that the precise location of a lightning strike could be influenced by the biota? Are snags or certain species more susceptible to lightning strikes? The answers to such questions are inconclusive, partly because of the difficulty of obtaining data. Often the ignition point can be determined if a fire is started, but the target of nonpyrogenic lightning is very difficult to ascertain. The available data are often imprecise or circumstantial, but are worthy of consideration nevertheless.

One of the earliest studies of lightning-caused forest fires was done by Barrows (1951), who found that the "first-ignited fuels" of 11,835 fires in the northern Rocky Mountains were, in decreasing frequency, snags 34%, duff on the ground 30%, wood on the ground 12%, green tree tops 10%, and miscellaneous other ignition points 14%. Barrows' data suggested that aerial fuels were ignited in 48% of the fires, which is higher than data reported by others (Plummer 1912) and which contradicts the theoretical conditions for a crown fire outlined by van Wagner (1977). Dead trees probably ignite more readily than live trees, and thus are more visible than lightning-struck live trees which do not ignite. Coulson et al. (1983) suggested that live, well-hydrated trees, with their roots in wet soil, should provide an electrical path of lower resistance than would a dead, dry tree, and Chapman (1950) reported that most strikes were to live trees in southeastern long-leaf pine forests, but the available information is inadequate for confident statements about the mechanisms determining the precise location.

Once started, some fires create another source of ignition, namely fire brands (Albini 1983; Pyne 1984; Stocks 1985). The "rain of fire brands" that may occur causes "spotting" downwind and can greatly magnify the effect of a single lightning strike. Furthermore, such ignitions facilitate fire spread across fuel discontinuities or other barriers.

The geography of lightning has been the subject of several studies, with the objective of determining whether some areas or forest types have a higher risk of strikes than others. Such work has involved the mapping of lightning fire zones, which Kourtz (1967) concluded do exist. One of the more detailed studies of this nature was that of Asleson and Fowler (Asleson 1975; Fowler and Asleson 1982; Fowler and Asleson 1984), who did a locational analysis of fires in northern Idaho. Although their data were based on fires, not strikes, their results suggested that grand fir forests, cedar-hemlock forests, and Engelmann spruce-subalpine fir forests were more likely to be ignited by lightning than would be expected based on the area each occupied. These results probably are due more to the fuel conditions that develop in these forests, or their topographic position, than to some inherent characteristic of the trees themselves. They found that, of 2087 fires, only 18% occurred below the upper two-thirds of the mountain (49% occurred on the upper one-third). Combining various factors, they concluded that the forests with the highest risk of lightning fire were those dominated by hemlock or Engelmann spruce and subalpine fir and occurring on the upper half of slopes perpendicular to the most common storm tracks. Similar studies were done by Hemstrom (1982) and Vankat (1985) in Mount Rainier National Park and Sequoia National Park, respectively.

Every lightning strike creates radio waves which can be used to locate the strike by triangulation. This principal has been adopted by industry (e.g., Lightning Location and Protection, Inc., Tucson, Arizona) for developing instrumentation capable of producing daily "lightning maps" for large areas. For example, this technology detected 786 ground strikes over Yellowstone National Park and vicinity during a 2.5-hour period on June 27, 1986; five fires were started but none continued to burn (D.G. Despain, personal communication). A more precise analysis of lightning strikes, as opposed to lightning-caused fires, may be possible as such data sets develop.

4.4 The Vegetation Mosaic in Nonforested Landscapes

Most studies in landscape ecology have tended to focus on regions dominated by forests, in part because many wildlands occur in regions of coniferous forest but also because boundaries between forests of different ages persist for many years and are easily detectable. The mosaic is very obvious, whether in a wilderness or an agrourban landscape, and the pattern clearly is not a simple function of environmental gradients.

Coniferous forests in particular lend themselves well to studies on the causes and patterns of the vegetation mosaic. Stand-replacing fires characterize such areas, along with significant episodes of parasitism, and, unlike many deciduous forests or chaparral, sprouting as a means of forest regeneration is a relatively rare phenomenon. Thus, the rate of recovery may be slower than in stands that develop by sprouting and the time available for occupancy by other species may be longer (Fig. 4.6). At the opposite extreme from coniferous forests are perennial grasslands, where very little structure develops aboveground, the fuel complex is more a function of annual shoot senescence than plant death or fragmentation, and a fire does not normally kill the plants. Recovery is rapid, being simply a matter of regrowth from live root crowns or rhizomes, and op-

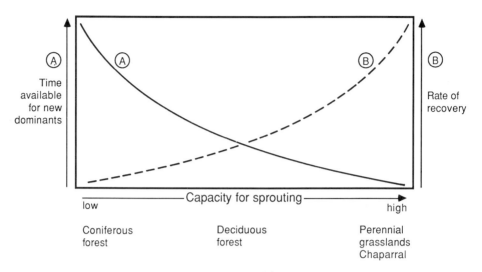

Figure 4.6. A generalized depiction of the possible relationships between the capacity for sprouting in various vegetation types, the rate of vegetation recovery after fire or some other major disturbances, and the time available after a disturbance for occupancy by new species. With some exceptions, coniferous forest mosaics may be more variable in space and time due to a lack of sprouting and relatively slow rates of patch recovery, as discussed in the text.

portunities for the invasion of pioneer species are limited. Some chapparal landscapes are very similar to grasslands in this regard, though fire frequency and the pattern of flammability development are quite different (Minnich 1983). The vegetation mosaic at the landscape scale in deserts or tundra may be affected by disturbances also, but environmental gradients probably are more influential under such rigorous climatic conditions.

Intermediate between grassland and coniferous forests are (1) deciduous forests, where some mortality may occur after fire; (2) mixed conifer-hardwood forests, where a large proportion of the stand may be killed resulting in a slower recovery; or (3) shrub-dominated grasslands such as sagebrush steppes where the shrubs may be killed by fire but most of the grasses and forbs are not. In the case of sagebrush steppe, recovery to the original composition is slower than grassland without sagebrush, but little opportunity exists for the invasion of other species (except for certain exotic species or under conditions of heavy grazing).

Vegetation mosaics caused by historical perturbations certainly must exist in landscapes dominated by deciduous forests or grasslands, where fires or other disturbances also have limits to their spread and where shifts in species composition occur at various scales including that of the landscape. Regrettably, the extensive tracts of deciduous forest, grassland, and shrubland required for such studies are difficult to find, if indeed they still exist at all.

4.5 Summary

The vegetation mosaic in wilderness landscapes at any particular time is determined by a broad range of factors including environmental gradients, continuing climatic change that may foster gradual changes in species composition, elements of chance with regard to location of ignitions and barriers to fire spread, and small scale disturbances that may increase the probability of larger scale conflagrations. In this chapter I have reviewed several mechanisms whereby biotic factors interacting with physical factors (even nonpyrogenic lightning) may enhance the probability of a fire occurring sooner (or later) than it might otherwise. The mechanisms seem intuitive and logical, though many must still be viewed as hypotheses and their importance will vary with vegetation type and position in the landscape.

Although one of the major challenges for landscape ecologists interested in wildlands is to deal simultaneously with the environmental and disturbance variables that cause the nature of the vegetation mosaic, an even greater challenge is understanding the ecological significance of the mosaic. What changes occur if the mosaic becomes homogenized due to fire suppression, or if the mosaic is modified as a result of timber harvesting or prescribed burning? What is the effect of having many small, contrasting patches instead of fewer large, contrasting patches? Answers to such questions are inherently interesting because wilderness landscapes were the setting for the evolution of most organisms, but also because one of the most conspicuous and universal effects of land management activities is to change vegetation mosaics. More efficient resource utilization and conservation could result when the importance and dynamics of the mosaic are understood as well as the dynamics of the patches.

Some interesting research on the significance of patch juxtaposition has already been done. For example, various studies have shown how fire spread may be retarded by a patchy vegetation mosaic (e.g., Minnich 1983; Heinselman 1985), and watershed hydrologists in the Rocky Mountains can now prescribe vegetation mosaics that maximize streamflow (Leaf 1975). Optimal mosaics for certain wildlife species also can be prescribed in some areas (Thomas et al. 1976; Harris 1984). Less information is available on the importance of the mosaic to, for example, (1) the preservation of biotic diversity; (2) maintaining a certain level of primary or secondary productivity; (3) the spread of weeds and the probability of insect epidemics or other disturbances; and (4) nutrient fluxes that could affect site productivity or streamwater quality. The results of wildland studies may be different from those for agrourban landscapes, but the insights gained could be useful to land managers everywhere.

Acknowledgments

Discussions with William H. Romme, William A. Reiners, and Don G. Despain have been very helpful, and I am grateful to the following persons for reviewing an earlier version of the manuscript: James K. Brown, Don G. Despain, Jerry

F. Franklin, Robert I. Gara, E.A. Johnson, David A. MacLean, Steward T.A. Pickett, William H. Romme, Monica G. Turner, and Peter S. White.

References

Agee, J.K. 1981. Initial effects of prescribed fire in a climax *Pinus contorta* forest: Crater Lake National Park. National Park Service Report, CPSU/UW 81-3. College of Forest Resources, University of Washington, Seattle, WA.

Albini, F.A. 1983. Potential spotting distance from wind-driven surface fires. USDA, Forest Service, Research Paper INT-309.

Albini, F.A. 1984. Wildland fires. *Am. Scientist* 72:590–597.

Alexander, M.E., Hawksworth, F.G. 1975. Wildland fires and dwarf mistletoes: A literature review of ecology and prescribed burning. USDA, Forest Service, General Technical Report RM-14.

Amman, G.D. 1977. The role of the mountain pine beetle in lodgepole pine ecosystems: Impact on succession, pp. 3–18. *In* W.J. Mattson (ed.), *The Role of Arthropods in Forest Ecosystems*. Springer-Verlag, New York.

Amman, G.D. 1978. Biology, ecology, and causes of outbreaks of the Mountain pine beetle in lodgepole pine forests, pp. 39–53. *In* D.L. Kibbee, A.A. Berryman, G.D. Amman, and R.W. Stark (eds.), *Theory and Practice of Mountain Pine Beetle Management in Lodgepole Pine Forests*. Forest, Wildlife, and Range Experiment Station, University of Idaho, Moscow, ID.

Andrews, P.L. 1986. BEHAVE: Fire behavior prediction and fuel modeling system— Burn subsystem, Part 1. USDA, Forest Service, General Technical Report INT-194.

Arno, S.F. 1980. Forest fire history in the northern Rockies. *J. Forestry* 78:460–465.

Arno, S.F., Sneck, K.M. 1977. A method for determining fire history in coniferous forests of the mountain west. USDA, Forest Service, Research Paper INT-202.

Asleson, D.O. 1975. Locational analysis of lightning-caused forest fires. Master's thesis, Department of Geography, University of Wyoming, Laramie, WY.

Barrows, J.S. 1951. Forest fires in the northern Rocky Mountains. USDA, Forest Service, Station Paper RM-28.

Berryman, A.A. 1972. Resistance of conifers to invasion by bark beetle-fungus associations. *BioSci.* 22:598–602.

Bormann, F.H., Likens, G.E. 1981. *Pattern and process in a forested ecosystem*. Springer-Verlag, New York.

Botkin, D.B. 1980. A grandfather clock down the staircase: Stability and disturbance in natural ecosystems, pp. 1–10. *In* R.H. Waring (ed.), *Forests: Fresh Perspectives From Ecosystem Analysis*. Proceedings of the 40th Annual Biology Colloquium, Oregon State University Press, Corvallis, OR.

Brown, J.K. 1975. Fire cycles and community dynamics in lodgepole pine forests, pp. 429–456. *In* D.M. Baumgartner (ed.), *Management of Lodgepole Pine Ecosystems*. Washington State University Cooperative Extension Service, Pullman, WA.

Brown, J.K., Oberheu, R.D., Johnston, C.M. 1982. Handbook for inventorying surface fuels and biomass in the interior west. USDA, Forest Service, General Technical Report INT-129.

Burgess, R.L., Sharpe, D.M. (eds.) 1981. *Forest Island Dynamics in Man-Dominated Landscapes*. Springer-Verlag, New York.

Cates, R.G., Redak, R., Henderson, C.B. 1983. Natural product defensive chemistry of Douglas fir, western spruce budworm success, and forest management practices. *Z. Angew. Entomol.* 96:173–182.

Chandler, C., Cheney, P., Thomas, P., Trabaud, L., Williams, D. 1983. *Fire in Forestry*, Vol. 1. Forest Fire Behavior and Effects. Wiley & Sons, New York.

Chapman, H.H. 1950. Lightning in the longleaf. *J. Am. Forestry* 56:10–11.

Christiansen, N.L. 1985. Shrubland fire regimes and their evolutionary consequences, pp. 85–100. *In* S.T.A. Pickett and P.S. White (eds.), *The Ecology of Natural Disturbance and Patch Dynamics*. Academic Press, New York.

Coulson, R.N., Hennier, P.B., Flamm, R.O., Rykiel, E.J., Hu, C., Payne, T.L. 1983. The role of lightning in the epidemiology of the southern pine beetle. *Z. Angew. Entomol.* 96:182–193.

Davis, M.B. 1981. Quaternary history and the stability of forest communities, pp. 132–153. *In* D.C. West, H.H. Shugart, and D.B. Botkin (eds.), *Forest Succession: Concepts and Application*. Springer-Verlag, New York.

Despain, D.G. 1985. Ecological implications of ignition sources in park and wilderness fire management programs, pp. 93–97. *In* J.E. Lotan, B.M. Kilgore, W.C. Fischer, and R.W. Mutch (technical coords.), *Proceedings: Symposium and Workshop on Wilderness Fire*. USDA, Forest Service, General Technical Report INT-182.

Despain, D.G., Sellers, R.E. 1977. Natural fire in Yellowstone National Park. *West. Wildlands* 4:20–24.

Dickman, A.W. 1984. Fire and *Phellinus weirii* in a mountain hemlock (*Tsuga mertensiana*) forest: Postfire succession and the persistence, distribution, and spread of a root-rotting fungus. Doctoral dissertation, University of Oregon, Eugene, OR.

Dinoor, A., Eshed, N. 1984. The role and importance of pathogens in natural plant communities. *Ann. Rev. Phytopath.* 22:443–466.

Edmunds, Jr., G.F., Alstad, D.N. 1978. Coevolution in insect herbivores and conifers. *Science* 199:941–945.

Ewel, J.E. 1980. Special issue on tropical succession. *Biotropica* 12 (suppl.):1.

Forman, R.T.T., Boerner, R.E. 1981. Fire frequency and the pine barrens of New Jersey. *Bull. Torrey Botan. Club* 108:34–50.

Forman, R.T.T., Godron, M. 1981. Patches and structural components for a landscape ecology. *BioSci.* 31:733–740.

Forman, R.T.T., Godron, M. 1986. *Landscape Ecology*. Wiley & Sons, New York.

Fowler, P.M., Asleson, D.O. 1982. Spatial properties of lightning-caused forest fires. *Physical Geog.* 3:180–189.

Fowler, P.M., Asleson, D.O. 1984. The location of lightning-caused wildland fires, Northern Idaho. *Physical Geog.* 5:240–252.

Franklin, J.F., Hemstrom, M.A. 1981. Aspects of succession in the coniferous forests of the Pacific, pp. 212–229. *In* D.C. West, H.H. Shugart, and D.B. Botkin (eds.), *Forest Succession: Concepts and Application*. Springer-Verlag, New York.

Furyaev, V.V., Wein, R.W., MacLean, D.A. 1983. Fire influences in *Abies*-dominated forests. pp. 221–234. *In* R.W. Wein and D.A. MacLean (eds.), *The Role of Fire in Northern Circumpolar Ecosystems*. Wiley & Sons, London.

Gara, R.I., Littke, W.R., Agee, J.K., Geiszler, D.R., Stuart, J.D., Driver, C.H. 1985. Influence of fires, fungi, and mountain pine beetles on development of a lodgepole pine forest in south-central Oregon, pp. 153–162. *In* D.M. Baumgartner, R.G. Krebill, J.T. Arnott, and G.F. Weetman (eds.), *Lodgepole Pine: The Species and its Management*. Washinton State University Cooperative Extension Service, Pullman, WA.

Gara, R.I., Agee, J.K., Littke, W.R., Geiszler, D.R., 1986. Fire wounds and beetle scars. *J. Forestry* 84:47–50.

Geiszler, D.R., Gara, R.I., Driver, C.H., Gallucci, V.F., Martin, R.E. 1980. Fire, fungi, and beetle influences on a lodgepole pine ecosystem in south-central Oregon. *Oecologia* 46:239–243.

Gruell, G.E. 1985. Indian fires in the interior west: A widespread influence, pp. 68–74. *In* J.E. Lotan, B.M. Kilgore, W.C. Fischer, and R.W. Mutch (technical coords.), *Proceedings: Symposium and Workshop on Wilderness Fire*. USDA, Forest Service, General Technical Report INT-182.

Habeck, J.R. 1974. Forests, fuels and fire in the Selway-Bitterroot Wilderness, Idaho, pp. 305–353. *In: Proceedings, Tall Timbers Fire Ecology Conference*, Missoula, MT.

Habeck, J.R. 1985. Impact of fire suppression on forest succession and fuel accumulations in long-fire-interval wilderness habitat types, pp. 110–118. *In* J.E. Lotan, B.M. Kilgore, W.C. Fischer, and R.W. Mutch (technical coords.), *Proceedings: Symposium and Workshop on Wilderness Fire.* USDA, Forest Service, General Technical Report INT-182.

Harris, L.D. 1984. *The Fragmented Forest: Island Biogeography Theory and the Preservation of Biotic Diversity.* University of Chicago Press, Chicago.

Hawksworth, F.G. 1975. Dwarf mistletoe and its role in lodgepole pine ecosystems, pp. 342–358. *In* D.M. Baumgartner (ed.), *Management of Lodgepole Pine Ecosystems.* Washington State University Cooperative Extension Service, Pullman, WA.

Hawksworth, F.G., Lister, C.K., Cahill, D.B., 1983. Phloem thickness in lodgepole pine: Its relationship to dwarf mistletoe and Mountain pine beetle. *Environ. Entomo.* 12:1447–1448.

Heinselman, M.L. 1973. Fire in the virgin forests of the Boundary Waters Canoe Area, Minnesota. *Quatern. Res.* 3: 329–382.

Heinselman, M.L. 1981a. Fire and succession in the conifer forests of Northern North America, pp. 374–405. *In* D.C. West, H.H. Shugart, and D. B. Botkin (eds.), *Forest Succession: Concepts and Application.* Springer-Verlag, New York.

Heinselman, M.L. 1981b. Fire intensity and frequency as factors in the distribution and structure of northern ecosystems, pp. 7–57. *In* H.A. Mooney, T.M. Bonnicksen, N.L. Christensen, J.E. Lotan, and W.A. Reiners (eds.), *Fire Regimes and Ecosystem Properties.* United States Department of Agriculture, Forest Service, General Technical Report WO-26.

Heinselman, M.L. 1985. Fire regimes and management options in ecosystems with large high-intensity fires, pp. 101–109. *In* J.E. Lotan, B.M. Kilgore, W.C. Fischer, and R.W. Mutch (technical coords.), *Proceedings: Symposium and Workshop on Wilderness Fire.* USDA, Forest Service, General Technical Report INT-182.

Hemstrom, M.A. 1979. A recent disturbance history of forest ecosystems at Mount Rainier National Park. Doctoral dissertation, Oregon State University, Corvallis, OR.

Hett, J. 1971. Landscape changes in east Tennessee and a simulation model which describes these changes for three counties. Ecological Sciences Division Publication 414 (ORNL-IBP-71-8), Oak Ridge National Laboratory, Oak Ridge, TN.

Johnson, P.C., Denton, R.E. 1975. Outbreaks of western spruce budworm in the American northern Rocky Mountains from 1922–1971. U.S.D.A., Forest Service, General Technical Report INT-20.

Johnson, W.C. 1977. A mathematical model of forest succession and land use for the North Carolina piedmont. *Bull. Torrey Botan. Club* 104:334–346.

Kessell, S.R. 1976a. Gradient modeling: A new approach to fire modeling and wilderness resource management. *J. Environ. Management* 1:39–48.

Kessell, S.R. 1976b. Wildland inventories and fire modeling by gradient analysis in Glacier National Park, pp. 115–162. *In: Proceedings, Tall Timber Fire Ecology Conference,* Missoula, Montana. Tall Timbers Research Station, Tallahasse, FL.

Kessell, S.R. 1979. *Gradient Modeling: Resource and Fire Management.* Springer-Verlag, New York.

Kilgore, B.M. 1981. Fire in ecosystem distribution and structure: Western forests and scrublands, pp. 58–89. *In* H.A. Mooney, T.M. Bonnicksen, N.L. Christensen, J.E. Lotan, and W.A. Reiners (eds.), *Fire Regimes and Ecosystem Properties.* United States Department of Agriculture, Forest Service, General Technical Report WO-26.

Komarek, Sr., E.V. 1964. The natural history of lightning, pp. 139–183. *In: Proceedings, Third Annual Tall Timbers Fire Ecology Conference,* Tallahassee, FL.

Kourtz, P. 1967. Lightning behavior and lightning fires in Canadian forests. Department of Forestry and Rural Development, Ottawa, Canada. Departmental publ. No. 1179.

Kozlowski, T.T., Ahlgren, C.E. (eds.) 1974. *Fire and Ecosystems.* Academic Press, New York.

Krawielitzki, V.S., Vit'e, J.P., Sturm, U., Francke, W. 1983. Interactions between resin flow and subcortically feeding Coleoptera. *Z. Angew. Entomol.* 96:140–146.

Leaf, C.F. 1975. Watershed management in the Rocky Mountain subalpine zone: The status of our knowledge. USDA, Forest Service, Research Paper RM-137.

Leopold, A.S. 1933. *Game Management.* Scribners, New York.

Lotan, J.E., Kilgore, B.M., Fischer, W.C., Mutch, R.W. (technical coords.), 1985a. *Proceedings: Symposium and Workshop on Wilderness Fire.* USDA, Forest Service, General Technical Report INT-182.

Lotan, J.E., Brown, J.K., Neuenschwander, L.F. 1985b. Role of fire in lodgepole pine forests, pp. 133–152. *In* D.M. Baumgartner, R.G. Krebill, J.T. Arnott, and G.F. Weetman (eds.), *Lodgepole Pine: The Species and its Management.* Washington State University Cooperative Extension Service, Pullman, WA.

Matson, P.A., Waring, R.H. 1984. Effects of nutrient and nitrogen limitation on mountain hemlock: susceptibility to laminated root rot. *Ecology* 65:1517–1524.

McCune, B. 1983. Fire frequency reduced by two orders of magnitude in the Bitterroot Canyon, Montana. *Can. J. Forest Res.* 13:212–218.

McDonald, G.I. 1981. Differential defoliation of neighboring Douglas fir trees by western spruce budworm. USDA, Forest Service, Research Note INT-306.

Minnich, R.A. 1983. Fire mosaics in southern California and northern Baja California. *Science* 219:1287–1294.

Mooney, H.A., Bonnicksen, T.M., Christensen, N.L., Lotan, J.E., Reiners, W.A. (eds.) 1981. *Fire Regimes and Ecosystem Properties.* United States Department of Agriculture, Forest Service, General Technical Report WO-26.

Mutch, R.W. 1970. Wildland fires and ecosystems—a hypothesis. *Ecology* 51:1046–1051.

Peterman, R.M. 1978. The ecological role of mountain pine beetle in lodgepole pine forests, pp. 16–26. *In* D.L. Kibbee, A.A. Berryman, G.D. Amman, and R.W. Stark (eds.), *Theory and Practice of Mountain Pine Beetle Management in Lodgepole Pine Forests.* Forest, Wildlife, and Range Experiment Station, University of Idaho, Moscow, ID.

Pickett, S.T.A. 1976. Succession: An evolutionary interpretation. *Am. Naturalist* 110:107–119.

Pickett, S.T.A., Thompson, J.N. 1978. Patch dynamics and the design of nature preserves. *Biol. Conserv.* 13:27–37.

Pickett, S.T.A., White, P.S. 1985. Patch dynamics: A synthesis, pp. 371–384. *In* S.T.A. Pickett and P.S. White (eds.), *The Ecology of Natural Disturbance and Patch Dynamics.* Academic Press, New York.

Plummer, F.G. 1912. Lightning in relation to forest fires. United States Department of Agriculture, Forest Service, Bulletin No. 111.

Pyne, S.J. 1984. *Introduction to Wildland Fire.* Wiley & Sons, New York.

Raffa, K.F., Berryman, A.A. 1982. Physiological differences between lodgepole pines resistant and susceptible to the Mountain pine beetle and associated microorganisms. *Environ. Entomol.* 11:486–492.

Reiners, W.A., Lang, G.E. 1979. Vegetational patterns and processes in the balsam fir zone, White Mountains, New Hampshire. *Ecology* 60:403–417.

Romme, W.H. 1982. Fire history and landscape diversity in Yellowstone National Park. *Ecolog. Monogr.* 52:199–221.

Romme, W.H., Knight, D.H. 1981. Fire frequency and subalpine forest succession along a topographic gradient in Wyoming. *Ecology* 62:319–326.

Romme, W.H., Knight, D.H. 1982. Landscape diversity: The concept applied to Yellowstone Park. *BioSci.* 32:664–670.

Romme, W.H., Knight, D.H., Yavitt, J.B. 1986. Mountain pine beetle outbreaks in the Central Rocky Mountains: Effects on primary productivity. *Am. Naturalist* 127:484–494.

Rothermel, R.C. 1983. How to predict the spread and intensity of forest and range fires. USDA, Forest Service, General Technical Report INT-143.

Rowe, J.S. 1979. Large fires in the large landscapes of the north, pp. 8–32. *In: Proceedings, Symposium on Fire Management in the Northern Environment*. USDI, Bureau of Land Management, Alaska State Office.

Royama, T. 1984. Population dynamics of the spruce budworm Choristoneura fumiferana. *Ecolog. Monogr.* 54:429–462.

Rundel, P.W. 1981. Structural and chemical components of flammability, pp. 183–207. *In* H.A. Mooney, T.M. Bonnicksen, N.L. Christensen, J.E. Lotan, and W.A. Reiners (eds.), *Fire Regimes and Ecosystem Properties*. United States Department of Agriculture, Forest Service, General Technical Report WO-26.

Schowalter, T.D., Coulson, R.N., Crossley, Jr., D.A. 1981. Role of southern pine beetle and fire in maintenance of structure and function of the southeastern coniferous forest. *Environ. Entomol.* 10:821–825.

Schowalter, T.D. 1985. Adaptations of insects to disturbance, pp. 235–252. *In* S.T.A. Pickett and P.S. White (eds.), *The Ecology of Natural Disturbance and Patch Dynamics*. Academic Press, New York.

Schmid, J.M., Hinds, T.E. 1974. Development of spruce-fir stands following spruce beetle outbreaks. USDA, Forest Service, Research Paper RM-131.

Shugart, H.H., Seagle, S.W. 1985. Modeling forest landscapes and the role of disturbance in ecosystems and communities, pp. 353–368. *In* S.T.A. Pickett and P.S. White (eds.), *The Ecology of Natural Disturbance and Patch Dynamics*. Academic Press, New York.

Sprugel, D.G. 1976. Dynamic structure of wave-regenerated Abies balsamea forests in the northeastern United States. *J. Ecology* 64:889–911.

Stocks, B.J. 1985. Forest fire behavior in spruce budworm-killed balsam fir, pp. 188–199. *In* C.J. Sanders, R.W. Stark, E.J. Mullins, and J. Murphy (eds.), *Recent Advances in Spruce Budworm Research*. Proceedings of the CANUSA Spruce Budworm Research Symposium, Bangor, ME, September 1984. Canadian Forest Service, Ottawa, Ontario.

Swain, A.M. 1980. Landscape patterns and forest history in the Boundary Waters Canoe Area, Minnesota. A pollen study from Hug Lake. *Ecology* 61:747–754.

Taylor, A.R. 1973. Ecological aspects of lightning in forests, pp. 455–482. *In: Proceedings, Tall Timbers Fire Ecology Conference*. Tall Timbers Research Station, Tallahassee, FL.

Thomas, J.W., Miller, R.J., Black, H., Rodiek, J.E., Maser, C. 1976. Guidelines for maintaining and enhancing wildlife habitat in forest management in the Blue Mountains of Oregon and Washington, pp. 452–476. *In: Transactions, 41st North American Wildlife and Natural Resources Conference*.

Tinnin, R.O. 1984. The effect of dwarf mistletoe on forest community ecology, pp. 117–122. *In* F.G. Hawksworth and R.F. Scharpf (eds.), *Biology of Dwarf Mistletoes: Proceedings of the Symposium*. United States Department of Agriculture, Forest Service, General Technical Report RM-111.

Tkacz, B.M., Schmitz, R.F. 1986. Association of an endemic mountain pine beetle population with lodgepole pine infected by *Armillaria* root disease in Utah. USDA, Forest Service, Research Note INT-353.

van Wagner, C.E. 1977. Conditions for the start and spread of crown fire. *Can. J. Forest Res.* 7:23–34.

van Wagner, C.E. 1978. Age class distribution and the forest fire cycle. *Can. J. Forest Res.* 8:220–227.

van Wagner, C.E. 1983. Fire behaviour in northern conifer forests and shrublands, pp. 65–80. *In* R.W. Wein and D.A. MacLean (eds.), *The Role of Fire in Northern Circumpolar Ecosystems*. Wiley & Sons, New York.

van Wagner, C.E. 1985. Does nature really care who starts the fire?, pp. 98–100. *In* J.E. Lotan, B.M. Kilgore, W.C. Fischer, and R.W. Mutch (technical coords.), *Proceedings: Symposium and Workshop on Wilderness Fire*. USDA, Forest Service, General Technical Report INT-182.

van Wagtendonk, J.W. 1985. Fire suppression effect on fuels and succession in short-fire-interval wilderness landscapes, pp. 119–126. *In* J.E. Lotan, B.M. Kilgore, W.C. Fischer, and R.W. Mutch (technical coords.), *Proceedings: Symposium and Workshop on Wilderness Fire*. USDA, Forest Service, General Technical Report INT-182.

Vankat, J.L. 1985. General patterns of lightning ignitions in Sequoia National Park, California, pp. 408–411. *In* J.E. Lotan, B.M. Kilgore, W.C. Fischer, and R.W. Mutch (technical coords.), *Proceedings: Symposium and Workshop on Wilderness Fire*. USDA, Forest Service, General Technical Report INT-182.

Waring, R.H., Pitman, G.B. 1983. Physiological stress in lodgepole pine as a precursor for Mountain pine beetle attack. *Zeitschrift fur Angewendte Entomologie* 96:265–270.

Waring, R.H., Pitman, G.B. 1985. Modifying lodgepole pine stands to change susceptibility to mountain pine beetle attack. *Ecology* 66:889–897.

Waring, R.H., Schlesinger, W.H. 1985. *Forest Ecosystems: Concepts and Management*. Academic Press, New York.

Webb, III, T. 1981. The past 11,000 years of vegetational change in eastern North America. *BioSci* 31:501–506.

Wein, R.W., MacLean, D.A. 1983. *The Role of Fire in Northern Circumpolar Ecosystems*. Wiley & Sons, New York.

Weinstein, D.A., Shugart, H.H. 1983. Ecological modeling of landscape dynamics, pp. 29–45. *In* H. Mooney and M. Godron (eds.), *Disturbance and Ecosystems*. Springer-Verlag, New York.

Wicker, E.F., Leaphart, C.D. 1976. Fire and dwarf mistletoe (Arceuthobium spp.) relationships in the northern Rocky Mountains, pp. 279–298. *In: Proceedings, 1974 Tall Timbers Fire Ecology Conference*. Tall Timbers Research Station, Tallahassee, FL.

Wright, Jr., H.E. 1974. Landscape development, forest fires, and wilderness management. *Science* 186:487–495.

Wright, Jr., H.E., Heinselman, M.L. 1973. The ecological role of fire in natural coniferous forests of western and northern North America. *Quatern. Res.* 3:319–328.

Zimmerman, G.T., Laven, R.D. 1984. Ecological interrelationships of dwarf mistletoe and fire in lodgepole pine forests, pp. 123–131. *In* F.G. Hawksworth and R.F. Scharpf (eds.), *Biology of Dwarf Mistletoes: Proceedings of a Symposium*. USDA, Forest Service, General Technical Report RM-111.

5. Fire, Grazing, and the Landscape Heterogeneity of a Georgia Barrier Island

Monica Goigel Turner and Susan P. Bratton

5.1 Introduction

Although the general role of ecological disturbance has received much recent attention (i.e., Lugo 1978; Barrett and Rosenberg 1981; Mooney and Godron 1983; Sousa 1984; Gerritsen and Patten 1985; Pickett and White 1985), the propagation and behavior of disturbance in landscape is not well understood (Risser et al. 1984; Wiens et al. 1985). Yet, it is the landscape level at which many environmental problems are managed. Many studies of disturbance are at too fine a scale for practical application. To gain a more complete hierarchical understanding of disturbance, we need to understand these processes at the larger scales.

At the landscape level, it is necessary to deal with heterogeneous systems. Previous studies have shown that responses to disturbance in homogeneous environments are complex, and at the landscape level this complexity may increase. However, the dynamics of heterogeneous environments have been, until recently, largely ignored by the ecological sciences (Risser et al. 1984).

When we move to the scale of the landscape, there are important questions regarding the spread of disturbance in landscapes that must be addressed. For example, how do disturbances move across the landscape? How does the heterogeneity of the landscape influence the propagation of a disturbance? Do effects vary with the type of disturbance or the type of landscape? How does the size or shape of patches influence the process?

This chapter explores three conceptual hypotheses which address some of these questions. We focus on how disturbances behave across a barrier island landscape, drawing examples from studies of disturbances, particularly fire and grazing, on Cumberland Island National Seashore, Georgia. The hypotheses are as follows:

1. Young, immature, resilient ecosystems may be energy sources for propagation of fire and grazing disturbances across the landscape. Mature, resistant, long-lived ecosystems may be recipients of severe impacts.
2. Increased landscape heterogeneity may increase the spread of disturbance between communities.
3. Multiple disturbances may have nonadditive effects and this tendency may be enhanced by increased landscape heterogeneity.

5.2   The Study Area

5.2.1   The Cumberland Island Landscape

Cumberland Island (30° 48′N, 81° 26′W) is the southernmost and largest of the Georgia sea islands (Fig. 5.1), measuring 28 km in length and 9 km across at its widest point. The major vegetation types in the Cumberland landscape are arranged in longitudinal bands that are primarily determined by physical factors such as tidal energy, salt spray, and topography. The island is composed of stable Pleistocene sediments, rarely subject to storm overwash and on which upland vegetation occurs, and more dynamic Holocene beaches, dunes, and marshes that are easily modified by major storm events. The interior forests are patchy, and patch size is generally small (Fig. 5.2). Some upland forests extend from the dunes to the sound on narrow parts of the island, whereas others are broken by sloughs, fresh water marshes, or large blocks of scrub. The island has been periodically farmed and logged since colonial times, and the patchiness of the upland forests is partly a result of these past disturbances.

The band of vegetation on the eastern, or ocean, side of the island consists of broad beaches backed by a row of foredunes. Interdune meadows and bayberry (*Myrica cerifera*) thickets separate these low, sparsely vegetated foredunes from higher secondary dunes on the north and south ends of the island. This habitat sequence is truncated in the middle of the island, where interdune meadows are poorly developed. On the western side of the island are extensive salt marshes composed primarily of monospecific stands of *Spartina alterniflora*.

The upland maritime forests stand between the secondary dunes and the salt marshes, and they are dominated by live oak (*Quercus virginiana*) and other hardwoods. As indicated earlier, these forests tend to be patchy. On the north end of the island, the oak forests occupy old dune ridges now separated by freshwater wetlands. The hydric sites support marshes primarily dominated by salt cordgrass (*Spartina bakeri*) or sawgrass (*Cladium jamaicense*). Behind this ridge and swale system lies more forest. Oaks predominate in some areas, whereas in others, pines, such as loblolly (*Pinus taeda*) or slash (*P. elliotii*),

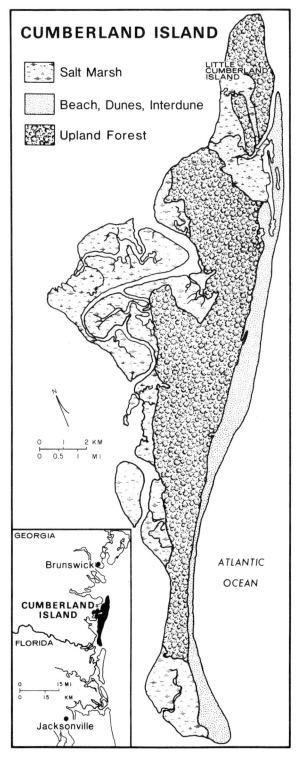

Figure 5.1. Map of Cumberland Island, Georgia, showing vegetation patterns of longitudinal bands.

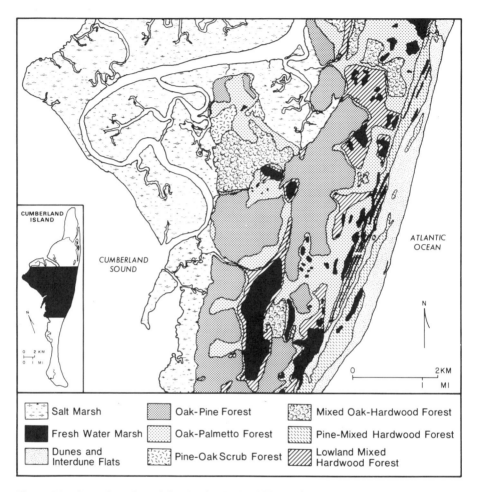

Figure 5.2. Central section of Cumberland Island illustrating the patchiness of the upland forests.

dominate or are mixed with hardwoods. Contiguous areas of pine and oak scrub are found in the north central part of the island and on Table Point. Small trees and ericaceous shrubs form a dense thicket on the most xeric interior soils.

The geology, soils, water resources, and fauna of Cumberland have been described by Hillestad et al. (1975). Designated as a National Seashore in 1972, the federal boundaries encompass approximately 14,500 hectares, about 9400 hectares of which is upland and marshes.

5.2.2 The Disturbances: Grazing and Fire

Physical forces or animals that move materials or energy in the landscape have been termed vectors (Forman 1981); Wiens et al. (1985) distinguished between biotic and abiotic vectors. Grazing and fire, the disturbances we will use to explore our hypotheses, represent these two types of vector. They are further distinguished by their origins. Disturbances can originate from either outside the ecosystem or within it (Bormann and Likens 1971); that is, the perturbation may be foreign to the system, or it may be from within but applied at an excessive level (Barrett et al. 1976). Fire is an abiotic disturbance that originates externally and is a point source perturbation which can then spread across the landscape. Grazing, in contrast, is a biotic disturbance that originates internally and is a diffuse source of disturbance. Animals may be present across the landscape, but as their numbers increase their impacts spread from preferred, or focal, habitats into others.

During the last century horses (*Equus caballus*), cattle (*Bos bovine*), and hogs (*Sus scrofa*) have ranged free on the island. The U.S. National Park Service removed the cattle in 1974 and has greatly reduced the hog population via trapping. The horse population, however, remains unmanaged and presently includes 180 animals (Finley 1985). The native white-tailed deer population (*Odocoileus virginianus*) has also reached a high density following release from hunting and removal of predators. Intensive grazing has depleted forest understories and reduced interdune and high marsh vegetation to the height of a mowed lawn.

The most important source of large fires on Cumberland Island is lightning, and the fires tend to occur during dry summers. At least 1300 hectares of Cumberland have burned since 1900, much of this area more than once, and fire rotations of 20 to 30 years have been documented (Turner 1984). The size of the burns varies with community type. Fires in scrub are large, hot, and frequently burn into other vegetation types. Fires in mature pine stands or in freshwater marshes may also spread into surrounding areas, but do not support as intense a fire as the scrub.

Fire and grazing are important factors determining the landscape mosaic on Cumberland. We suggest that although they operate quite differently, there are some general tendencies governing their behavior in the landscape, and that this behavior is dependent upon landscape heterogeneity.

5.3 Spread of Disturbance Across the Landscape: Hypotheses

5.3.1 Resilient Ecosystems May Be Energy Sources for Disturbance

Hypothesis 1: Young, immature, resilient ecosystems may be energy sources for propagation of grazing and fire disturbances across the landscape. Mature, resistant, long-lived ecosystems may be recipients of severe impacts.

Although the processes of combustion and ingestion differ, both fire and grazing are disturbances that "consume" biomass as they proceed. Biomass is required as a fuel for the propagation of these disturbances. In the landscape, one ecosystem or patch may serve as a fuel source for the disturbances, encouraging further spread into other patches or the surrounding matrix. We suggest a pattern in this process.

5.3.1.1 Grazing by Horses and Deer

The feral horses on Cumberland depend primarily on salt marshes and interdune grassland for forage (Table 5.1). Salt marshes comprise 34% of the island and represent the single largest habitat available to the horses. Lenarz (1983) found that horses on the south end of the island spent 51.7% of their total foraging time in salt marsh and estimated the genus *Spartina* to be 25.5% of the horses' diet. Although the horses obtain a large portion of their energy from the salt marsh, they also use other ecosystems, such as the interdune meadows. Thus, between-patch interactions and transfers occur. Horses move between prime feeding areas and freshwater sources, traversing other communities, especially the maritime forest, in the process. Territorial conflicts may initiate movements of a kilometer or more, and frequently result in movements of several hundred meters (S. Bratton, field observations). Horses leave the marshes during high tides and to escape biting insects, using the forests as a high ground refuge from the tides, as shelter and, to a limited extent, for foraging. On the north end, where salt marshes are farther from the interdune meadows, horses may spend more time in the forest and may consume forest genera such as oak (*Quercus* spp.) and Spanish moss (*Tillandsia usneoides*).

Without the marshes and grasslands, the forests of Cumberland Island would probably not support a viable horse herd. The dense forest understories are dominated by unpalatable saw palmetto (*Serenoa repens*), and the more open communities are low in productivity. With the energy supplied by the marsh, however, horses use the forests, removing or trampling vegetation that probably includes young oaks, and depositing feces. Large piles of horse manure in buckthorn-sabal palmetto thickets adjoining Dungeness marsh indicate that the forest may be gaining nutrients (S. Bratton, field observations).

It is clear that the horses move across boundaries in the landscape. What effects do they have on the ecosystems they use? Horse grazing and trampling in the salt marshes drastically decrease the standing stock (Table 5.2) and net aboveground primary productivity of *Spartina alterniflora* (Turner 1985, 1987). Experimental studies indicate that two components of grazing, trampling and removal of biomass, both have significant, independent effects on the marsh. There is no shift in species composition, however, and *Spartina alterniflora* continues to dominate. After release from horse grazing, the marsh shows fairly

Table 5.1. Proportion of Foraging Time Spent in Each Habitat for Feral Horses on Cumberland Island, Georgia[a]

Habitat	Percentage of Observations	Percentage of Habitat
Salt marsh	51.7	34.1
Grasslands at marsh edge	22.5	2.6
Interdune meadows	15.3	6.3
Lawns	8.1	1.2
Foredunes	2.1	8.6
Other	0.3	47.2

[a] Lenarz 1983.

Table 5.2. Peak Biomass of *Spartina alterniflora* in Grazed and Ungrazed Plots, Cumberland Island, Georgia[a]

	Ungrazed		Grazed	Biomass Removed by Horses (%)
Plot A				
1983	120		52	57
1984	364		9	98
1985	528		8	98
Plot B				
1983	68		14	79
1984	232		7	97
1985	140		5	96
Plot C				
1983	388		292	25
1984	440		180	60
1985	472		28	94
Plot D				
1983	344	ns	382	0
1984	382	ns	372	0
1985	472		228	52

[a] Units are dry weight (g/m^2). Grazed and ungrazed are significantly different ($P<0.05$, Student's t-test, N = 8), except where indicated by ns. Study plots are arranged from nearest to the uplands (A) to farther in the marsh (D).

high resilience: *Spartina* begins to recover immediately (Turner 1985, 1987) and may regain the standing crop of ungrazed areas in three years.

To assess grazing effects in the forests, horse exclosures were erected on Cumberland in 1985, but no data from these are available yet. However, data from another barrier island with similar communities suggest processes that may also occur on Cumberland. Bratton and Davison (1985) compared understory species composition in maritime forest at Buxton Woods, Cape Hatteras National Seashore, NC, between 1937, when livestock were removed, and 1984. In 1937, there were practically no oak seedlings in the understory, and pine seedlings were prevalent. Since the removal of feral animals, oak seedlings have increased tremendously (Table 5.3). Decrease in grazing not

Table 5.3. Ratio of Percent Cover of Oak Seedlings to Pine Seedlings, Buxton Woods, Cape Hatteras[a]

Habitat Type	1937[b]	1984
Pine forest	0.10	7.11
Hardwood forest	0.48	31.50
Scrub	n/a	108.00

[a] Bratton and Davison 1985.
[b] Feral animals were removed in 1937.

only released the understories, but along with changes in fire frequencies, resulted in a decrease in pine seedlings and the edge zone species, persimmon (*Diospyros virginiana*) and beauty berry (*Callicarpa americana*), whereas oaks (*Quercus* spp.) and bays (*Persea* spp.) increased (Table 5.3). This suggests that the presence of feral ungulates may change the composition of the maritime forest from oak dominance to pine.

The overall pattern of horse impacts is illustrated in Fig. 5.3A. Horses obtain most of their energy from the salt marshes and grasslands, ecosystems which are both highly disturbed and resilient. However, the horses may have substantial effects on the maritime forest, which alone is not sufficient to maintain the herd. This suggests that on Cumberland, the marshes and grasslands are the energy sources for a perturbation which may change forest community structure and species composition.

Deer intensively graze the interdune meadows, open grassy fields, and upper edges of the salt marsh, but do not use the low marsh as do the horses. Deer will also browse almost all broad-leaved woody species on the island. Census data (unpublished) indicate that the density of deer may exceed 0.5 per hectare in the upland areas.

The overall pattern of the deer disturbance is shown in Fig. 5.3B. The disturbance results from an increase in the population density, causing impacts to spread from focal habitats into nonfocal patches within the landscape. Deer browsing is selective, and may slowly modify species composition in the forests.

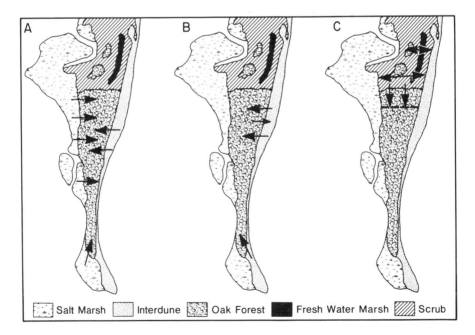

Figure 5.3. Energy flow of the three major disturbances across the barrier island landscape: (A) feral horses; (B) deer; (C) fire. Note that all affect the maritime forest.

Severe browse impacts may shift understory seedling composition from oak to pine (Bratton 1980), thus effecting changes similar to those caused by feral animals.

If grassland habitats were unavailable, deer would probably persist in low numbers in the forest. Thus, unlike the horses, a viable population might be maintained with decreased heterogeneity. However, the intensity of deer activity in the forests is increased by the availability of other habitats. The grassy, open communities appear to provide a source of energy that increases the grazing disturbance in the forest.

5.3.1.2 Fire

Fires will be initiated only in communities that contain adequate amounts of fuel. On Cumberland, fires generally originate in the scrub or freshwater marsh ecosystems where the vertical continuity of fuel encourages combustion of most available plant material. The fires are very hot and may spread completely across a patch of scrub or marsh once ignited. Scrub communities are probably "pyric disclimaxes" (Davison 1984) that accumulate fuels quickly and are prone to burning during periodic coastal droughts (Webber 1935).

Depending on weather, flammability, and the availability of fuel, fire may spread outward from its source. On Cumberland, fires may significantly affect communities that border the scrub or marshes. For example, a fire in the scrub will burn the edges of a mature oak forest, whereas the oak forest alone would not provide conditions conducive to burning.

During the summer of 1981, a severe fire burned from pine-oak scrub into surrounding communities on Cumberland. At the oak forest edge, oak trees within 3 m of the edge were usually killed by the fire, but the intensity of the fire decreased as it moved farther under the canopy. The effects of the fire did not extend into the forest as far as grazing did, usually penetrating only about 30 m. Patches of pine forest were more vulnerable to the fire, probably because of greater accumulations of litter. Patches of hardwoods surrounded by pine litter were killed completely and few sprouted. Live oak stands with basal areas greater than 13 m^2 suffered much less mortality than those with low basal area, such as those at the edge of the scrub (Davison 1984). A large stand of oaks will retard a fire and may protect small patches of other communities within its boundaries. However, small patches of mature oak forest probably cannot survive within larger patches of scrub.

The scrub tends to replace itself after fire (Webber 1935; Veno 1976; Davison 1984), showing high resilience. Small freshwater marshes are also fire tolerant and regenerate more quickly than the surrounding pine lands. The oak forest, in contrast, is fire resistant, and recovery in burned areas is slow. Thus, the presence of scrub patches within the landscape creates a focus for fire, which can move into the landscape and convert forest to scrub. Scrub vegetation is perpetuated by periodic destructive fires that will not permit further succession (Davison 1984).

The locations of the scrub and oak forests may be partially controlled by topographic factors, yet fires from the scrub encroach on the mature oak. The boundary between these two communities may be determined by a relationship between edaphic conditions, fire intensity, and potential stand size. However, the physical disturbance of vegetation removal by fire can directly alter boundary locations, thereby influencing landscape patterns (Wiens et al. 1985). The scrub community may not be more productive than a live oak forest, but more biomass is consumed by fire, making it the energetic source of this disturbance in mature oak (Fig. 5.3C).

5.3.1.3 Discussion

Fire and grazing disturbances on Cumberland Island appear to be driven by resilient patches within the landscape, yet significantly affect resistant patches from which little energy is obtained. Why might this occur?

Vertical stratification differs between the source and recipient habitats on Cumberland. The source communities (marsh, meadow, scrub) are low and continuous, concentrating their productivity within the first few meters above the ground. Much of this energy is readily available to grazers and is susceptible to fire. In contrast, the recipient, live oak forest, has two or three vertical strata, and the mature trees possess characteristics that make them resistant to fire and grazing. Most of the internal energy of the forest system is unavailable to horses, deer, or fire, but the live oak forest is vulnerable to both fire and grazing in its understory. This vulnerability is important because if no young oaks survive, the forest will ultimately disappear.

Regeneration processes also differ between the source and recipient habitats, both in terms of temporal scale and reproductive strategy. The marsh, meadow, and scrub communities are composed of species that recover quickly from fire and grazing. The predisturbance structure of the community may be achieved again in a few years. The live oak forest, however, consists of long-lived species that recover very slowly. Depending on available light and other environmental factors, young hardwoods may take 15 to 20 years to grow out of the reach of grazers. Fire resistance in trees generally increases as trees grow taller and bark increases in thickness. The typical rotation for scrub fires is 10 to 20 years (Komarek 1974; Christensen 1981), suggesting they may recur before the oaks are protected.

The spatial arrangement of the habitats on Cumberland also contributes to the spread of the disturbances. The arrangement of the communities in longitudinal bands places different habitats in very close proximity across the width of the island. This increases the probability of a vector encountering and crossing a boundary. For example, grazers may move between preferred habitats easily, and in doing so traverse ecosystems that they might otherwise not use. Furthermore, patches in the upland forests tend to be small, thus having a high edge/area ratio and a relatively high proportion of boundary that can be crossed.

Forman and Godron (1984, 1986) suggested that in an undisturbed landscape, horizontal landscape structure tends progressively toward homogeneity, and vertical structure becomes higher and more heterogeneous. Disturbances usually rapidly produce opposite patterns of heterogeneity. The Cumberland landscape appears to fit this pattern. In the absence of disturbance, the maritime oak forest would presumably form a homogeneous horizontal cover across the upland portion of the island, with a concomitant increase in vertical stratification.

Wiens et al. (1985) presented a conceptual framework for the study of boundary dynamics in landscapes. They suggested that the differences in between-patch and within-patch fluxes of energy, materials, or information create directional imbalances in the fluxes. Through such inequities, boundaries may shift in location or become blurred or sharpened through time. Our results support this concept, as well as Margalef's (1963) principle that when two systems of different maturity meet along a boundary that allows an exchange, energy flows toward the more mature system. However, our results are contrary to the second part of Margalef's principle, that the energy transfer favors the expansion of the mature patch into the immature patch. Instead, we observe an expansion of the less mature ecosystem. However, Margalef was specifically addressing stages of successional maturity, and not differences in the resistance and resilience of the patches. For example, although a pine forest and oak forest have differing successional maturity, the salt marsh and oak forest are both mature ecosystems. However, the marsh and oak forests differ in their abilities to resist or recover from specific disturbances.

The combination of differences in available energy, structure, stratification, resistance, and resilience of landscape components may have a substantial influence on both the spread of disturbance in the Cumberland landscape and the resulting landscape pattern. Transference of disturbance from one system to another may be a major determinant of long-term landscape dynamics.

5.3.2 Heterogeneity Increases the Spread of Disturbance

Hypothesis 2: Increased landscape heterogeneity may increase the spread of some disturbances.

We have observed that fire may burn from scrub into the surrounding mature forest matrix, and that grazing animals may move from preferred patches to less preferred habitats in the landscape. Given that fire and grazing do indeed spread from one habitat to another, what is the effect of changing the spatial heterogeneity of the habitats on the spread of these disturbances? As Wiens et al. (1985) note, it is part of conventional wisdom in ecology that disturbances are more likely to spread across a homogeneous area than one that is heterogeneous. We suggest, however, that the opposite also occurs. The movement of a disturbance from one habitat to another is amenable to modeling, and to address this issue, we developed a simple model to simulate different levels of heterogeneity in an idealized landscape.

The model assumed a square landscape that included two habitats, each of which comprised 50% of the landscape. A disturbance (i.e., fire on Cumberland Island) could travel unidirectionally for a fixed distance across the boundary from source habitat A (i.e., scrub) to recipient habitat B (i.e., oak forest). The landscape could then be divided into any even square number of boxes (4, 9, 16, etc.) with habitats A and B alternating systematically. The total area of the landscape, the distance the disturbance could travel, and the heterogeneity of the landscape (the number of boxes) were varied. The amount of habitat B that was impacted by the disturbance was then determined.

Simulations were run using fire and grazing disturbances, and estimates of the distance the disturbances travel were based on data and field observations. Results illustrate that increased heterogeneity enhanced the spread of these disturbances. Simulating the spread of fire from the scrub community to the live oak forest, increased patchiness in a 10,000 m² landscape resulted in 100% of the oak forest burned when 25 squares were present (Table 5.4). In a 1,000,000 m² landscape, 100% of the oak forest was burned when the landscape was di-

Table 5.4. Results of the Model Simulating the Spread of Disturbance with Increasing Landscape Heterogeneity

Disturbance:	Fire	Source Habitat:	Scrub
Landscape Area:	10,000 m²	Recipient Habitat:	Oak
Disturbance Travels:	20 m		

No. squares	Size per square (m²)	% Habitat affected
4	2,500	64
9	1,111	96
16	625	96
25	400	100

Disturbance:	Fire	Source Habitat:	Scrub
Landscape Area:	1,000,000 m²	Recipient Habitat:	Oak
Disturbance Travels:	20 m		

No. squares	Size per square (m²)	% Habitat affected
4	250,000	8
9	111,111	15
16	62,500	22
36	27,777	36
64	15,625	48
100	10,000	60
400	2,500	94
625	1,600	100

Disturbance:	Grazing	Source Habitat:	Marsh
Landscape Area:	100,000,000 m²	Recipient Habitat:	Oak
Disturbance Travels:	500 m		

No. squares	Size per square (m²)	% Habitat affected
4	25,000,000	19
9	11,111,111	36
16	6,250,000	51
36	2,777,777	64
64	1,562,500	91
100	1,000,000	99
121	826,446	100

vided into 625 squares, each of which had an area of 1600 m². In both cases
the curve of percent oak forest affected versus the number of squares was
hyperbolic; the percent affected increased with the amount of edge (total area
of each habitat remained constant).

Similar results were observed for grazing effects. Assume an area of
100,000,000 m² consisting of marsh and live oak forest with grazers moving 500
m from the marsh to the forest (Table 5.4). With four squares, only 19% of the
forest is affected; 16 squares results in impacts in 51% of the forest. By 64
squares, more than 90% is impacted, and this then increases gradually to 100%.

In the Pacific northwest, Franklin and Forman (1986) found a similar relationship between clearcut and remnant patches of forest and the spread of
windthrow into the remnant patches. As the size of the remnant patches decreased after a checkerboard pattern, susceptibility to windthrow increased.

In general, the trends of these simple simulations indicate that with large
areas considerable subdivision of the landscape is required before the recipient
habitat is severely impacted. With small areas, few subdivisions will have large
effects. Although this model is simplistic, it illustrates some potential effects
of landscape heterogeneity on the spread of disturbance. Further refinement
could be added to simulate random arrangements of patches, variable areas of
each habitat, different shapes and sizes of patches, and even accounting for a
gradient in the intensity of the disturbance as it spreads.

These results are consistent with a principle of landscape ecology proposed
by Forman and Godron (1984, 1986), that the flow of heat energy and biomass
across boundaries between patches, corridors, and matrix increases as landscape
heterogeneity increases. The fire and grazing disturbances on Cumberland represent exactly these types of flows across a landscape. However, there are
other cases in which increased heterogeneity may retard the spread of disturbance. Pine bark beetle attacks fall into this category, along with the spread
of pests in agroecosystems, some wildfire perpetuation, the spread of Dutch
elm disease, and erosional patterns (Risser et al. 1984). Thus, the geometry of
the landscape alone does not determine the nature of the relationship between
heterogeneity and disturbance. Ecological characteristics and differences between patches will determine this. The relative importance of geometry and
biological factors in determining the spread of disturbance remains to be determined.

Whether landscape heterogeneity enhances or retards the spread of the disturbance depends on the interaction among several factors, including (1) the
type and scale of the disturbance, (2) the spatial arrangement of components
within the landscape, (3) the characteristics of those components and their
permeability to the disturbance, and (4) the similarity of adjacent patches with
respect to the disturbance.

5.3.3 Nonadditive Effects of Multiple Disturbances

Hypothesis 3: Multiple disturbances may have nonadditive effects on the landscape, and this tendency may be enhanced by increased landscape heterogeneity.

Thus far, only fire and grazing disturbances acting separately in the landscape have been addressed. However, landscapes are responding to complexes of disturbances that occur at different spatial and temporal scales. It has been suggested that multiple disturbances may have nonadditive effects in ecosystems (Lugo 1978; White and Pickett 1985; Schowalter 1985; Turner 1985). We suggest that multiple disturbances may also have nonadditive effects in the landscape, and that this tendency may be enhanced by increased heterogeneity.

When more than one perturbation occurs simultaneously, there are several alternative responses that may be observed. First, there may be no response observable because the perturbations act in opposite directions and cancel each other. Or, the response may be additive, being the sum of the responses to the individual perturbations. Finally, the response may be synergistic (greater than additive) or antagonistic (less than additive). Measuring the effects of both combined and single factors in the same study is critical to identify the interactions of multiple factors.

Nonadditive effects of multiple disturbances were documented in a single ecosystem on Cumberland Island by Turner (1985). The impacts of multiple disturbances (clipping, trampling, and fire) on the salt marsh were studied to determine whether some combinations of disturbance could elicit nonadditive responses from the vegetation. Responses of standing stocks of *Spartina* to combinations of burning plus clipping and burning plus trampling were significantly antagonistic: the response to these perturbations in combination was not as severe as expected based on the disturbances applied singly (Turner 1985). The study demonstrated that multiple disturbances could indeed elicit nonadditive responses from components of an ecosystem. We hypothesize that such interactions may also occur across a landscape, and that increased heterogeneity of the landscape may enhance the opportunity for synergistic or antagonistic effects.

Fire and grazing may have opposite effects on the landscape over relatively short time scales. For example, fire may decrease the heterogeneity of the landscape, burning through small patches of maritime forest and increasing the amount of scrub. Grazing, in contrast, may decrease fire frequency by removing available fuel, reducing the effect of fire. Thus, in the short term, the disturbances are antagonistic and their combined effect is less than expected. However, grazing may encourage the development of pine stands in openings in the upland forests by selective foraging on oaks. These pine stands will burn more readily than the oak forest. By encouraging the occurrence of patches of pine forest, grazing will enhance the spread of fire and reduce any potential oak recovery. Thus, over the long term, grazing may interact synergistically with fire to enhance the spread of the fire-resilient scrub. This interaction may reduce the forest matrix to remnant patches which are then susceptible to extinction.

As discussed earlier, increased heterogeneity of the landscape tends to increase the spread of fire and grazing disturbances. Similarly, increased heterogeneity provides more opportunities for nonadditive interactions among disturbances to occur. It is likely that a landscape consisting of fewer large patches will have fewer synergistic disturbances than a landscape that contains many small patches.

To fully understand the dynamics occurring within a landscape, the entire disturbance regime must be considered along with potential nonadditive interactions among disturbances. On Cumberland Island, for example, the vegetation mosaic is also influenced by feral pigs, the activities of human visitors to the island, and storms. Major storms may dramatically alter the landscape in a very short time, perhaps less than a day. The effects of these disturbances on shifting boundaries and transfers across the landscape must also be assessed to fully understand the landscape dynamics.

5.4 Conclusion

On Cumberland Island, the heterogeneity of the landscape strongly influences the behavior of grazing and fire disturbances. In turn, the boundaries between components of the landscape may change as a result of disturbance. Our analysis suggests some additional questions for research. It is unclear, for example, to what extent the community boundaries are a function of environmental conditions and to what extent they are determined by disturbance dynamics. The environmental factors may determine the absolute limits of a community type, whereas the disturbance regime controls the arrangement within that ultimate constraint. Furthermore, is antagonism among disturbances a common phenomenon that acts as a stabilizing force in the landscape?

We also suggested that one ecosystem or patch may serve as a fuel source for the disturbances, encouraging further spread into other patches or the surrounding matrix. In general, does small patch size and the prevalence of resilient or successional patches indicate large energy transfers across the landscape? Conversely, does large patch size and the dominance of resistant or mature communities indicate lower energy transfer?

This study also suggests that management of particular patches (i.e., for wildlife) or disturbances (i.e., fire) in the landscape may have a substantial effect on community boundaries and thus the structure of the landscape. The importance of assessing management actions in the context of the whole landscape is underscored.

In a heterogeneous landscape such as Cumberland Island, differences in structure, stratification, available energy, resistance, and resilience between landscape components may have a substantial influence on the spread of disturbance and the resulting landscape pattern. Transference of disturbance from one system to another may be a major determinant of long-term landscape dynamics. To fully understand the dynamics occurring within a landscape, the entire disturbance regime must be considered along with potential nonadditive interactions among disturbances. Whether landscape heterogeneity enhances or retards the spread of disturbance will depend on the interaction of several factors, including (1) the type and scale of the disturbance, (2) the spatial arrangement of components within the landscape, (3) the characteristics of those components and their permeability to disturbance, and (4) the similarity of adjacent patches with respect to the disturbance.

Acknowledgments

We thank Jerry F. Franklin, Frank B. Golley, Eugene P. Odum, William E. Odum, and Paul G. Risser for their critical comments and suggestions on this manuscript. Research discussed in this paper was funded by the U.S. National Park Service, Cooperative Park Studies Unit, University of Georgia.

References

Barrett, G.W., Van Dyne, G.M., Odum, E.P. 1976. Stress ecology. *BioSci.* 26(3):192–194.
Barrett, G.W., Rosenberg, R. (eds.) 1981. *Stress Effects on Natural Ecosystems.* Wiley & Sons, New York.
Bormann, F.B., Likens, G.E. 1971. The ecosystem concept and the rational management of natural resources. *Yale Scientific Mag.* April 2–8.
Bratton, S.P. 1980. The impacts of white-tailed deer on the vegetation of Cades Cove, Great Smoky Mountains National Park. *Proc. Ann. Conf. SE Assoc. Fish & Wildlife Agencies* 33:305–312.
Bratton, S.P., Davison, K.L. 1985. The disturbance history of Buxton Woods, Cape Haters, NC. *CPSU Tech. Rep. 16,* Institute of Ecology, University of Georgia, Athens, GA.
Christensen, N.L. 1981. Fire regimes in southeastern ecosystems: Southeastern United States, pp. 112–136. *In Fire Regimes and Ecosystem Properties.* Proc. Conf. USDA For. Serv. Gen. Tech. Rep. WO.26, Honolulu, HI.
Davison, K.L. 1984. Vegetation response and regrowth after fire on Cumberland Island National Seashore, GA. Master's Thesis, University of Georgia, Athens, GA.
Finley, M. 1985. Structure of the feral horse population, 1985: Cumberland Island National Seashore, Camden County, GA. *Natl. Park Serv. Coop. Park Studies Unit, Univ. Georgia Tech. Rep. 17.*
Forman, R.T.T. 1981. Interaction among landscape elements: A core of landscape ecology, pp. 35–48. *In Proc. Int. Congr. Neth. Soc. Landscape Ecol.,* Veldhovenm 1981. Pudoc, Wageningen, 1981.
Forman, R.T.T., Godron, M. 1984. Landscape ecology principles and landscape function, pp. 4–16. *In* J. Brandt and P. Agger (eds.), *Methodology in Landscape Ecological Research and Planning,* Vol. V. Roskilde University Centre Book Company, GeoRuc, Roskilde, Denmark.
Forman, R.T.T., Godron, M. 1986. *Landscape Ecology.* Wiley & Sons, New York.
Franklin, J.F., Forman, R.T.T., 1986. Influences of patch clearcutting on potential for catastrophic disturbance in coniferous forest landscapes of the Pacific Northwest. *Symposium Presentation: The Role of Landscape Heterogeneity in the Spread of Disturbance,* University of Georgia, Athens, GA.
Gerritsen, J., Patten, B.C. 1985. State space system theory of ecological disturbance. *Ecol. Modelling* 29:383–397.
Hillestad, H.O., Bozeman, J.R., Johnson, A.S., Berisford, C.W., Richardson, J.L. 1975. The ecology of Cumberland Island National Seashore, Camden County, GA. *Report to the Natl. Park Service, Tech. Report Series No. 75–5.*
Komarek, E.V. 1974. Effects of fire on temperate forests and related ecosystems: Southeastern United States, pp. 251–277. *In* T.T. Kozlowski and C.E. Ahlgren (eds.), *Fire and Ecosystems.* Academic Press, New York.

Lenarz, M.S. 1983. Population size, movements, habitat preferences, and diet of the feral horses of Cumberland Island National Seashore. *Natl. Park Serv. Coop. Park Studies Unit, Univ. Georgia Tech. Rep. No. 3.*

Lugo, A.E. 1978. Stress and ecosystems, pp. 61–101. *In* J.W. Thorp and J.W. Gibbons (eds.), *Energy and Environmental Stress in Aquatic Ecosystems.* U.S. Dept. Energy Symp. Series CONF 77114.

Margalef, R. 1963. On certain unifying principles in ecology. *Am. Naturalist* 97:357–374.

Mooney, H.A., Godron, M. (eds.) 1983. *Disturbance and Ecosystems.* Springer-Verlag, New York.

Pickett, S.T.A., White, P.S. (eds.) 1985. *The Ecology of Natural Disturbance and Patch Dynamics.* Academic Press, New York.

Risser, P.G., Karr, J.R., Forman, R.T.T. 1984. Landscape ecology. Directions and approaches. Illinois Nat. Hist. Survey Special Publ. 2. Ill. Nat. Hist. Surv., Champaign.

Schowalter, T.D. 1985. Adaptations of insects to disturbance, pp. 235–252. *In* S.T.A. Pickett and P.S. White (eds.), *The Ecology of Natural Disturbance and Patch Dynamics.* Academic Press, New York.

Sousa, W.O. 1984. The role of disturbance in natural communities. *Ann. Rev. Ecol. Syst.* 15:353–391,

Turner, M,M.G. 1985. Ecological effects of multiple perturbations on a Georgia salt marsh. Doctoral dissertation, University of Georgia, Athens, GA.

Turner, M.G. 1987. Effects of feral horse grazing, clipping, trampling and burning on a Georgia salt marsh. *Estuaries* 10(1). In press.

Turner, S. 1984. The fire history of Cumberland Island National Seashore 1900–1983. *Natl. Park Serv. Coop. Park Studies Unit, Univ. Gerogia. Tech. Rep. 7.*

Veno, P.A. 1976. Successional relationships of five Florida plant communities. *Ecology* 57:498–508.

Webber, H.J. 1935. The Florida scrub, a fire fighting association. *Am. J. Bot.* 22:344–361.

White, P.S., Pickett, S.T.A. 1985. Natural disturbance and patch dynamics: An introduction, pp. 3–13. *In* S.T.A. Pickett and P.S. White (eds.), *The Ecology of Natural Disturbance and Patch Dynamics.* Academic Press, New York.

Wiens, J.A., Crawford, C.S., Gosz, J.R. 1985. Boundary dynamics: A conceptual framework for studying landscape ecosystems. *Oikos* 45:421–4627.

6. Disturbance by Beaver (*Castor canadensis* Kuhl) and Increased Landscape Heterogeneity

Marguerite Madden Remillard, Gerhard K. Gruendling, and Donald J. Bogucki

6.1 Introduction

A landscape is defined as an area having a common geomorphology, climate, and disturbance regime encompassing all types, frequencies, and intensities of disturbance through time (Mooney and Godron 1983; Forman and Godron 1986; and Risser, Chapter 1). Disturbances are discrete events in time which disrupt an ecosystem, community, or population structure and change resources, substrate availability, or the physical environment (Pickett and White 1985). The relationship between disturbance and heterogeneity in a landscape is complex and depends on the scale of the disturbance and important underlying environmental gradients. Disturbances may either increase or decrease heterogeneity (Denslow 1985), whereas landscape heterogeneity may enhance or inhibit the spread of disturbance (Risser et al. 1984).

Disturbances to sessile communities typically produce open spaces, change the availability of resources such as light, soil nutrients, and moisture and allow the establishment of new propagules (Denslow 1985). Disturbances thus create patches in the landscape. Patches are discrete communities embedded in an area of dissimilar community structure or composition, the matrix (Mooney and Godron 1983).

One type of patch, a spot disturbance patch, results from disturbance of a small area within a matrix, for example a blowdown in a forest or a small fire in a grassland. After the disturbance, succession proceeds until the patch becomes like the surrounding matrix. If the intensity or the frequency of the disturbance is high, the patch may remain semistable, differing significantly from the matrix (Forman and Godron 1981, 1986). Beaver create spot disturbance patches in a forest matrix which then undergo succession, thereby increasing landscape heterogeneity.

Beaver impound waterways and create ponds; dam construction normally being the first activity undertaken upon arrival at a new homesite (Seaton 1928; Shadle 1956; Rue 1964; Hodgdon 1978). Inundation dramatically alters the ecological community from a terrestrial to an aquatic system that will remain as long as the dams are maintained. When the dams are breached, the ponded area drains and may revert to a terrestrial environment. The site may later be reoccupied by beaver and reflooded (Johnson 1927; Neff 1957; Rue 1964).

Fluctuating water levels cause plant communities to successively replace one another based upon their tolerance of hydric conditions. Typical changes in streamside habitat above beaver dams and immediately surrounding the pond have been described by Johnson (1927), Wilde et al. (1950), and Knudson (1962). Common terms for abandoned beaver sites include beaver meadows and beaver flowages.

> Now, while the popular mind readily recognizes that the beaver meadow may be a logical successor to the beaver pond, it does not so easily grasp that there are other possibilities, or that the beaver meadow is not necessarily an end result in itself; that it may be but a step in the onward march of events; that it represents but a stage in an orderly process of succession in nature and is due in its turn to be followed by other conditions. Seeds of forest trees will find lodgement and growth in this fertile soil and ultimately the meadow will have been succeeded by the sheltering swamp or the wooded valley, as the case may be, . . . (Johnson 1927).

An investigation of succession related to beaver disturbance requires a record of past vegetation and beaver activity, such as historical aerial photography. Aerial photographs have been used to analyze temporal changes in vegetation (Kershaw 1964; Johnson 1969; Howard 1970; Kirby 1976; Barber et al. 1977), as well as active and abandoned beaver sites (Dickinson 1971b; Parsons and Brown 1978; Howard 1982; Gruendling et al. 1985).

The objectives of this investigation were: (1) to examine the feasibility of the use of historical aerial photographs to investigate plant community changes caused by beaver in the Adirondack State Park of New York, and (2) to determine whether stages in plant community succession in beaver patches exist and investigate spatial and temporal changes in vegetation related to beaver activity.

6.2   Methods

6.2.1   Study Sites

The Adirondack State Park of New York (Fig. 6.1) was created in 1892 and encompasses 2.2 million hectares of private and state-owned land. Forty percent of the Park is preserved as "forever wild." Elevation ranges from 29 m in the

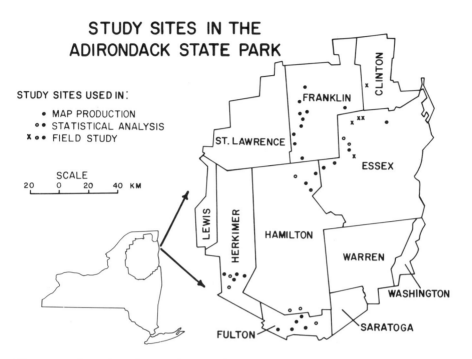

Figure 6.1. Location of study sites within the Adirondack State Park of New York.

northeast, to 1629 m in the central High Peaks region. Erosion has exposed ancient Precambrian substratum, with the oldest exposed rocks estimated to be more than 1 billion years old. The geomorphology of the Adirondacks is dominated by glacial landforms created by the Wisconsin ice sheet 12,000 years ago (Jamieson 1985; Terrie 1985).

The area is predominantly forested with tree species associated with glacial deposition type and resulting soils. *Picea rubens* (red spruce) and *Pinus strobus* (white pine) are found in mineral soils of glacial outwash. The lower mountain slopes and the peripheral plateau region are covered by unsorted till and mixed forests of deciduous and evergreen trees. Where deeper layers of organic soil have accumulated, the climax forest consists of northern hardwoods such as *Acer saccarum* (sugar maple) and *Fagus grandifolia* (beech), and *Tsuga canadensis* (hemlock) (Jamieson 1985).

Aerial photographs of the Adirondack State Park were examined to locate 39 beaver sites surrounded by forest and relatively isolated from agriculture, transportation routes, and residential development. A beaver site was defined as a pond or series of ponds used by a colony of beavers and separated from other beaver sites by an unaltered section of stream (Howard 1982; Howard and Larson 1985). All available historical aerial photography was used to map the vegetation. Eleven sites mapped at least once in each decade from 1940 to 1980 were selected for areal measurement of vegetation types and statistical analysis. Five of these sites were also selected as representative of patches in various stages of succession and suitable for field studies. All study sites are indicated in Fig. 6.1.

6.2.2 Acquisition of Aerial Photographs

The *Inventory of Aerial Photography and Other Remotely Sensed Imagery of New York State* (Map Information Unit 1979), was used to locate agencies possessing historical aerial coverage of the study sites (Table 6.1.) The film types of the photographs recorded between 1942 and 1981 were black and white panchromatic and black and white infrared in 23- × 23-cm positive-print format. Scales ranged from 1:15,000 to 1:33,000.

To document the current plant communities at five field sites, color and color infrared aerial photographs were taken on September 15, 1983. An aerial camera system designed by Bogucki and Gruendling (1978), with twin Hasselblad EL-500 cameras (f = 50 mm) was used. The cameras held two film types, Kodak Ektachrome MS Aerographic Film (2448 Estar Base) and Kodak Aerochrome Infrared Film (2443 Estar Base), producing color and color infrared 70-mm positive transparencies, respectively. A flight altitude of 1000 m above ground level resulted in a nominal scale of 1:20,000. The photographs were recorded between 11:00 AM and 2:00 PM to reduce shadow in the mountainous area.

6.2.3 Mapping Procedure

Vegetation maps were made of 39 beaver sites for each year of available aerial photography recorded between 1942 and 1983. Due to limitations in the quality of historical aerial photographs (often being faded, folded, torn, and of poor resolution and low contrast), the small scale, and black and white film type, only general vegetation cover types could be interpreted. The following were identified: hardwood, softwood, shrub, emergent, open water, and dead trees.

For each beaver site, vegetation in the area immediately affected by fluctuating water levels was mapped and designated the central zone or (00) boundary. To record vegetation in the surrounding matrix that were unaffected by flooding, zones up to 100 m (01 boundary) and 200 m (02 boundary) from the edge of the central zone were also mapped (Fig. 6.2). A Bausch and Lomb

Table 6.1. Sources of Aerial Photographs Used to Map Beaver Colonies

Adirondack Park Agency, Raybrook, NY
Johnstown Real Property Tax Service, County of Fulton, Johnstown, NY
New York State Department of Environmental Conservation, Albany, NY
New York State Department of Environmental Conservation, Northville, NY
New York State Department of Transportation Mapping Unit, Albany, NY
State Board of Equalization and Assessment, Saranac Lake, NY
State University of New York at Plattsburgh, Plattsburgh, NY
Soil Conservation Service, County of Franklin, Malone, NY
United States Department of Agriculture, Johnstown, NY

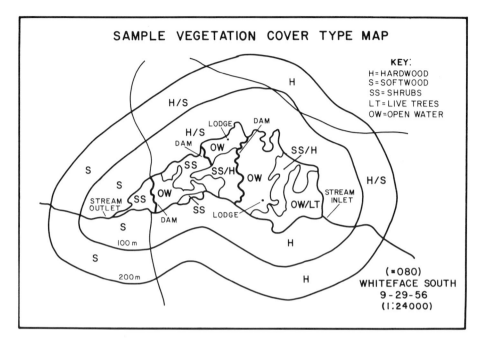

Figure 6.2. Vegetation map of a beaver patch.

Stereo Zoom Transfer Scope was used to enlarge the photographic images and produce vegetation cover type maps. Because of the tilt and relief displacement inherent in the aerial photographs, percent cover rather than actual area was obtained from the vegetation maps using a LASICO Digital Planimeter.

6.2.4 Field Studies

Field studies were necessary to verify interpretation of vegetation cover types from the aerial photographs. Vegetation analysis was conducted between July 11, 1983, and October 4, 1983. Circular quadrats were systematically located within the (00), (01), and (02) boundaries (Fig. 6.3). Concentric circular plots of (1, 80, and 200 m²) were established using a Rem Rod and BAF-10 Exact Factor Wedge Prism (Nyland and Remelle 1975) to sample herbaceous, shrub-seedling-sapling, and tree vegetation, respectively (Fig. 6.4). Herbaceous species were identified, and percent cover was visually estimated. Shrubs, saplings, and tree species were identified, the diameter at breast height was measured, and Importance Values based on relative dominance, density, and frequency were calculated.

6.2.5 Data Analysis

Data analysis consisted of two multivariate statistical techniques, discriminant analysis and Bray-Curtis polar ordination. Data were considered multivariate

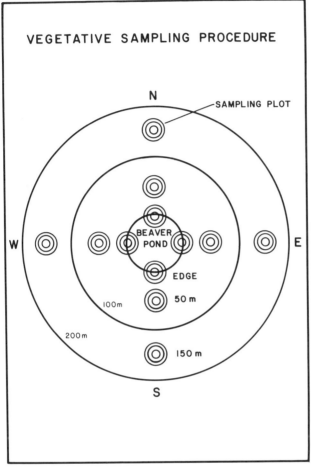

VEGETATIVE SAMPLING PROCEDURE

Figure 6.3. Field sampling plots systematically located around the former or existing beaver pond.

because each site was described by the abundances of 17 cover type combinations. Discriminant analysis was performed using the *Statistical Package for the Social Sciences (SPSS)* (Klecka 1975), and Bray-Curtis polar ordination by use of the Fortran program, ORDIFLEX, by Gauch (1977).

To perform discriminant analysis using SPSS, data were organized into cases of site per year per boundary. Each case was assigned a descriptor of its vegetation composition by summing percent cover for similar cover types into four general types or stages: open water, emergent, shrub, and mixed forest softwood/hardwood. Cases were then assigned the stage or stages occupying at least 70% of the site area. Discriminant analysis was performed to distinguish among boundaries (00), (01), and (02), and among stages based on discriminating variables of vegetation cover type.

Bray-Curtis polar ordination represents sample (or case) and species (or cover type) relationships in a low-dimensional space (Gauch 1982). Data were arranged

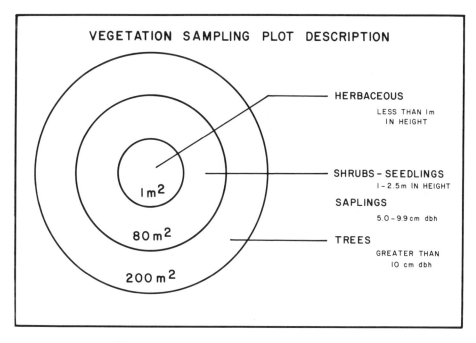

Figure 6.4. Description of field sampling plots.

in a cases by cover types matrix of cover type abundances: 107 cases by 17 cover types in size. To describe the vegetation in cases graphed in the ordination space, each case was identified by the most abundant cover type. Cases described by cover types with less than three occurrences were dropped from the ordination analysis to eliminate outlier samples and clarify the results. The original data matrix was reduced to 97 cases by 16 cover types. Cases were then compared with automatically selected endpoints based on their vegetation composition. Percent dissimilarity, which linearly weights species or cover type abundances, was used as a measure of comparison. The positions of cases relative to the endpoint cases were graphed in a two-dimensional ordination space.

To investigate the direction of successional changes in vegetation caused by beaver, the sequence of plant community changes or stage replacement in each site was examined on sequential aerial photographs. A diagram illustrating the direction and frequency of stage replacement for all sites was produced by drawing one arrow for each stage succeeded by another stage. Based on the dates between sequential aerial photographs, temporal aspects of succession related to beaver were obtained. The average number of years for sites to succeed from open water stages indicating beaver activity, through other stages of abandonment and back to open water stages was calculated as a cycle of activity-abandonment-activity. The average of these cycles indicates the average time between disturbances by beaver or the disturbance return interval.

6.3 Succession in Beaver Patches

Discriminant analysis distinguished among beaver site boundaries and among stages representative of plant communities for each year of available historical aerial photography. As only the first of two discriminant functions generated was statistically significant (Table 6.2), vegetation of the central (00) boundary was significantly different from vegetation in the surrounding (01) and (02) boundaries. Cases plotted along the first two discriminant function continuums illustrate the separation of the (00) boundary from the (01) and (02) boundaries (Fig. 6.5). Group centroids, the most typical location in the discriminant function space of a case from a group, are represented by an asterisk.

Cases were next discriminated by stage based on variables of percent cover for 17 vegetation cover types. Although seven canonical discriminant functions were produced, the sixth and seventh functions accounted for a minimum amount of the variance (Table 6.3). Five functions thus define six significant groups. A graph of cases and group centroids along the first two discriminant functions (Fig. 6.6), illustrates an apparent continuum of stages from emergent (top) to shrub (bottom) with intermediate stages of open water/emergent, open water, and open water/shrub. Although there is a forested stage designated softwood/hardwood, there are no transitional stages between softwood/hardwood and other stages.

Bray-Curtis polar ordination also indicated that emergent, shrub, shrub/emergent, and open water stages may be identified in a continuum of successional stages within the (00) boundary. Polar ordination produced a graph illustrating case positions relative to two endpoints within a low-dimensional ordination space. Four main clusters of cases are evident in Fig. 6.7: open water, emergent, shrub, and shrub/emergent. Cases clustering in the center of the graph appear to be transitional and again indicate a continuum of succession.

Plant species were identified in five field sites selected as representative of the various stages of plant community succession. Moose Pond was an active beaver site, with the (00) boundary consisting of open water and dead trees surrounded by red spruce, *Betula lutea* (yellow birch), and hemlock trees and saplings. Evidence of beaver activity included a large pond and lodge, well-maintained dams, fresh cuttings, and observed beaver. The Silver Lake site exemplified an abandoned beaver site in the emergent stage. Emergents consisted mainly of *Calamagrostis canadensis* (blue joint grass), *Carex lacustris* (sedge), *Sphagnum* spp. (Sphagnum moss), and *Iris versicolor* (blue flag). The shrub/emergent stage, Whiteface South, supported a mixture of *Alnus rugosa* (alder) and emergents such as *Juncus effusus* (rush), sedge, and grass species. A beaver site once abandoned and then recently reoccuppied, Franklin Falls, consisted mainly of flooded alder shrubs and open water.

The sequence of successional stages, recorded by each year of available historical aerial photography, was examined for spatial and temporal trends in plant community succession caused by beaver. Due to gaps in the sequence of historical aerial coverage of most sites, beaver activity and vegetation between the dates of available photographs is unrecorded and unaccounted for. Thus,

Table 6.2. Canonical Discriminant Functions for Discriminant Analysis by Boundary

Function	Eigenvalue	Percent of Variance	Cumulative Percent	Canonical Correlation	After Function	Wilk's Lambda	Chi-Square	Degrees of Freedom	Level of Significance
1	21.38	99.60	99.60	0.98	0	0.04	770.73	32	0.00
2	0.09	0.40	100.00	0.28	1	0.92	20.02	15	0.17

Table 6.3. Canonical Discriminant Functions for Discriminant Analysis by Stage

Function	Eigenvalue	Percent of Variance	Cumulative Percent	Canonical Correlation	After Function	Wilk's Lambda	Chi-Square	Degrees of Freedom	Level of Significance
1	99.29	81.93	81.93	0.99	0	0.00	2422.70	112	0.00
2	14.51	11.97	93.90	0.97	1	0.00	1321.40	90	0.00
3	6.46	5.33	99.23	0.93	2	0.06	666.16	70	0.00
4	0.60	0.50	99.73	0.61	3	0.46	185.82	52	0.00
5	0.21	0.17	99.90	0.41	4	0.74	73.13	36	0.00
6	0.07	0.06	99.96	0.26	5	0.89	28.45	22	0.16
7	0.05	0.04	100.00	0.22	6	0.95	11.47	10	0.32

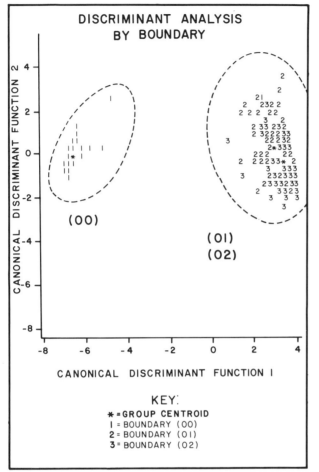

Figure 6.5. Plot of cases in discriminant analysis by boundary.

only general trends in the temporal and spatial aspects of stage replacement will be discussed.

Successional stage replacement related to beaver activity is multidirectional and nonlinear, as evidenced by arrows indicating the occurrence of one stage followed by another stage in the (00) boundary of all sites (Fig. 6.8). The most frequent sequence of successional stage replacement is from the open water stages (open water, open water/emergent, and open water/shrub), to emergent, to shrub/emergent, to shrub. Conversion back to open water stages is approximately equal in frequency from emergent, shrub/emergent, and shrub stages. Only two sites were observed on 1942 photographs as the original forest matrix, and no sites were observed to return to forest cover. The direction of stage replacement may reverse, stages in the sequence may be skipped, or stages may be repeated on sequential aerial photographs.

The beaver disturbance return interval was obtained by examining the dates

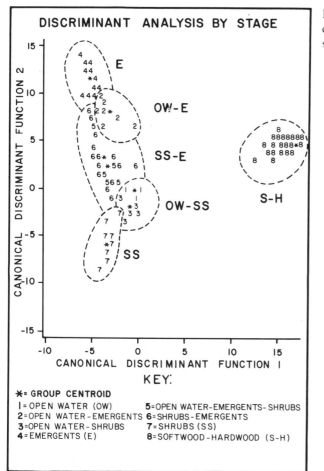

Figure 6.6. Plot of cases in discriminant analysis by stage.

of sequential aerial photographs and the successional stages of beaver sites. Because gaps exist of 1 to 10 years between available historical aerial photographs for beaver sites in the Adirondacks, only a general time scale for successional stage replacement was determined. Sites completed a cycle of stage replacement from open water stages, through other stages and back to open water stages in an average of 22 ± 8 years, from emergent stages to emergent stages in 19 ± 9 years, and from shrub to shrub stages in 18 ± 8 years. The beaver disturbance return interval is thus approximately 10 to 30 years.

6.4   Discussion of Beaver Disturbance

Although physical processes of disturbance such as fire and wind in landscapes have been studied extensively (Daubenmire 1968; Vogl 1974; Henry and Swan 1974; Sprugel 1976; Bormann and Likens 1979; Heinselman 1981; Forman and

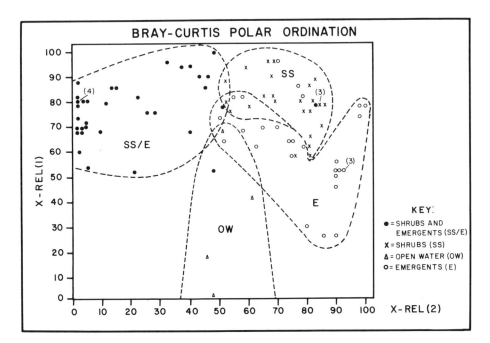

Figure 6.7. Plot of cases in Bray-Curtis polar ordination.

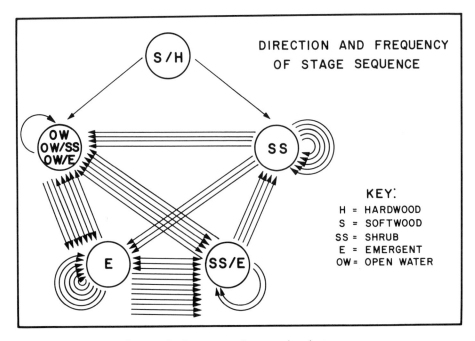

Figure 6.8. Sequence of successional stages.

Boerner 1981; Romme and Knight 1981; Romme 1982; and others), the importance of animals in a landscape has also been recognized. Animals as agents of disturbance include predators, herbivores, and animals with nonpredatory behaviors that inadvertently kill or displace other organisms (Golley et al. 1975; Connell and Slatyer 1977; Whittaker 1977; Sousa 1984). Mammals have an impact on the kinds of plants present in an ecosystem, their abundance, distribution, form, and reproduction. Small mammals within a system cause direct effects on biological components, on the rates of transfers between components, and on the physical environment (Golley et al. 1975).

Some examples of investigations addressing animals as major disturbances in a landscape are the following: cyclic insect outbreaks in plant communities (Blais 1954; Allen and Foster 1977; Peterman 1978), vertebrate grazers in grasslands (Curtis 1956; Dix 1959; Itow 1963; Harper 1969; Naveh and Kinski 1975; McNaughton 1979, 1983), badger in the tall-grass prairie (Platt 1975), boar in Appalachian beech forests (Bratton 1974, 1975), and feral horses in salt marshes (Turner 1985). Muskrats (*Ondatra zibethicus*) disturb marshes by harvesting aquatic vegetation and creating openings. Although muskrat are a disturbance factor in prairie pothole cyclic succession, drought is the major disturbance controlling water levels, succession, and the muskrat populations themselves (Weller and Spatcher 1965; Weller and Fredrickson 1974; van der Valk and Davis 1978; Weller 1981).

Beaver also affect the landscape (Johnson 1927; Ives 1942; Neff 1957; Chadbreck 1958; Knudson 1962; Champagne 1971; Dickinson 1971b; White 1979; Vogl 1981; Remillard 1984). According to Vogl (1981), "beavers in many North American locations with lakes, rivers and streams are classical examples of retrogressive agents whose destruction/disturbance checks and reverses plant succession and determines the stage of vegetational development."

Beaver may affect vegetation by harvesting trees, shrubs, and herbaceous plants for food and building materials, and by impounding waterways and flooding vegetation. Beaver cut trees and clear small areas in a forest; however, these clearings are not as distinct at the spatial scale of the aerial photographs as the openings caused by beaver dam construction and flooding. Beaver flooded areas are clearly observed on the aerial photographs as patches in a forested landscape. The size of patches created by beaver in a forest matrix is variable. Beaver flowages included in a survey of central Adirondack wetlands ranged from less than 0.4 to 24.3 hectares (Dickinson 1971a). Beaver patches used as field sites in this study varied in size from approximately 1.5 to 12 hectares and averaged 6.7 hectares.

Flooding by beaver changes a terrestrial system into an aquatic system. Environmental effects by beaver were described in early investigations by Radford (1907), Johnson (1927), and Wilde et al. (1950):

> The erection of a beaver dam introduces certain changes in the environment and in the established mode of life. The killing of trees by impounding water is not the most serious but the most conspicuous detrimental outcome. Removal of the dam by dynamiting or other means involves another drastic modification of conditions, and produces still further consequences. The entire cycle affects the po

sition of the ground water table, fertility of the soil, rate of forest growth and the status of game within the flowage area, as well as on adjacent lands (Wilde et al. 1950).

More recently, Pritchard and Hall (1971), Hodkinson (1975a, 1975b, 1975c), Naiman and Melillo (1984), and Naiman et al. (1984) have examined ecological effects of beaver activity. Forman and Godron (1986) describe beaver as a stream corridor animal that "plays an unusually significant role when present." Beaver dam construction and destruction upon abandonment of the beaver site perpetuates a state of vegetation flux that may increase habitat diversity and species diversity (Forman and Godron 1986).

Vegetation within beaver patches changes with fluctuating water levels associated with cyclic abandonment and reoccupation of beaver sites. Rainfall variation is not considered the major controlling factor of plant succession or beaver populations in Adirondack beaver patches. Beaver compensate for fluctuating water levels by building dams across streams and maintaining stable ponds (Seaton 1928; Shadle 1956; Rue 1964; Hodgdon 1978; Weller 1981). Most streams in the Adirondack Mountains flow throughout the growing season due to collective runoff in high terrain and relatively constant precipitation. The Western Adirondacks receive an average annual precipitation of more than 127 cm with relatively uniform distribution throughout the year and rare severe droughts (Ruffner 1980). The effect of drought on succession in beaver patches is thus minimized, and plant community changes are believed to be caused by fluctuating water levels related to beaver activity.

Plant communities successively replace one another based on their tolerance of hydric conditions. These stages include emergent, open water/emergent, open water, open water/shrub, shrub, and mixed forest softwood/hardwood. The creation of open water, emergent macrophyte, and shrub patches in a forest matrix increases landscape heterogeneity.

Successional changes in plant communities within beaver patches are nonlinear and multidirectional. Most frequently, beaver patches succeeded from open water, to emergent, to shrub/emergent, to shrub stages, and back to open water from emergent, shrub/emergent, and shrub stages. Stages may also perpetuate themselves, the sequence may skip stages, and the direction of stage replacement may reverse.

Classic concepts of linear succession may not apply when disturbance recurs at regular intervals (Noble and Slatyer 1980). Nonlinear sequences of plant community replacement have been suggested by Egler (1954), Horn (1976), Connell and Slatyer (1977), and Noble and Slatyer (1980). These schemes of succession also allow that the replacement sequence can restart at one of several positions after a disturbance. "In some systems there may be several tracks that the system takes depending on the time when the system is affected (frequency) and the intensity of the so-called disturbance" (Vogl 1981).

The variable sequence and direction of beaver patch plant community succession may be due to lack of documentation for existing plant communities between years of available historical aerial coverage, unique physical and vegetative characteristics of the beaver sites, or to the frequency of disturbance dependent upon abandonment and reoccupation of sites by beaver. The return interval of beaver disturbance was approximately 10 to 30 years. The 20-year range in the return interval cycle is expected because many factors influence abandonment and reoccupation of beaver sites. Beaver ponds may remain active and water levels stable for as long as 8 to 10 years (Knudson 1962). Site abandonment on 29 sites studied by Hodgdon (1978) was not related to any single factor, although a severe shortage of hardwood food resources was apparent in 62% of the locations. Most authors attribute site abandonment to severe depletion of the food supply, and reoccupation of abandoned sites to expanding populations or to the regrowth of the food supply following previous occupation (Hall 1971).

In the Adirondacks, trapping and food supply depletion are the most probable causes for site abandonment. The reoccupation of beaver sites in the shrub, shrub/emergent, or emergent stages depends on the cause of site abandonment. If beaver have been removed by trapping before they have depleted the food supply, new beaver may occupy the site during the emergent stage. If former beaver occupants have harvested all of the available food and building resources, a longer period is needed to replenish the site. Socioeconomic factors influencing beaver trapping pressure may thus indirectly determine the directional and temporal aspects of plant community succession in beaver patches.

The succession of stages among open water, emergent, and shrub stages without a return to the mixed forest softwood/hardwood stage (Fig. 6.8), as well as the separation of the softwood/hardwood stage from the continuum of other stages (Fig. 6.6), suggest that once a forest is disturbed by beaver the patch will succeed to the shrub stage and will not return to the original forest matrix. The patch would then be considered a semistable spot disturbance patch remaining distinct from the surrounding matrix structure and composition as described by Forman and Godron (1981, 1986). This may be possible if flooding has dramatically altered the environment of the patch, for example, by raising the ground water level. Alder, the dominant species in shrub stages of beaver patch succession, commonly occupy poorly drained soils and border stream banks in thickets where surface drainage is slow and the ground water level is close to the surface during part of the growing season (Gill and Healy 1974).

It should be noted, however, that the historical aerial photographs were recorded over only 40 years, and a longer period of investigation may be necessary to observe the return of forest cover in beaver patches. Disturbance by beaver may also be too frequent for the sites to return to a forest. The high density of beaver in the Adirondacks due to relatively low value of the pelts and reduced trapping pressure may favor reoccupation of shrub patches by beaver.

Finally, beaver colony sites that have returned to the forest matrix may not have been recognized on the aerial photographs as former beaver sites. A longer period of investigation or conclusive evidence that presently forested areas were former beaver patches must be obtained before conclusions concerning plant community succession in patches without further disturbance by beaver may be made.

6.5   Conclusions

Based on the results of this investigation on beaver disturbance and plant community succession in the Adirondack State Park of New York, the following conclusions may be made:

1. Beaver create spot disturbance patches in a forest matrix by constructing dams, impounding streams, and flooding existing vegetation. After abandonment and drawdown of the pond, plant communities change according to their tolerance of hydric conditions. The creation of open water, emergent macrophyte, and shrub patches in a forest increases landscape heterogeneity.
2. Historical aerial photographs may be used to record and quantify past plant community changes in patches created by beaver.
3. Plant community succession in beaver patches is a continuum with the following six phases or stages identified by discriminant analysis: emergent, open water/emergent, shrub/emergent, open water/shrub, shrub, and mixed forest softwood/hardwood. Bray-Curtis polar ordination identified four stages in beaver patch succession: open water, emergent, shrub/emergent, and shrub.
4. The pattern of plant community succession in beaver patches is multidirectional and nonlinear with patches repeatedly being occupied, abandoned, and reoccupied by beaver. The beaver disturbance return interval is approximately 10 to 30 years, with equal probability of shrub, shrub/emergent, and emergent stages being disturbed by beaver.

Acknowledgments

The authors would like to express their gratitude to the various agencies listed in Table 6.1 for use of aerial photographs; to Eleanor Madden, Rachel Krotz, and Andrew Deecher for assistance in map measurement and fieldwork; and to Dr. Kenneth Adams for advice in statistical methods. The review and comments on the manuscript by Dr. Monica Turner and two anonymous reviewers were also greatly appreciated.

References

Allen, C.T., Foster, D.E. 1977. Seasonal food habits of the desert termite on a shortgrass prairie in west Texas. Res. Highlights Noxious Bush and Weed Control, Range Wildlife Management, 8:49, Texas Tech University, Lubbock, TX.

Barber, E., Bogucki, D.J., Gruendling, G.K., Madden, M. 1977. Historical land use changes and impacts in Lake Champlain wetlands. Lake Champlain Basin Study, New England River Basins Commission, Burlington, VT.

Blais, J.R. 1954. The recurrence of spruce budworm infestations in the past century in the Lac Seul area of northwestern Ontario. *Ecology* 35:62–71.

Bogucki, D.J., Gruendling, G.K. 1978. Remote sensing to identify, assess and predict ecological impact on Lake Champlain wetlands. State University of New York, Plattsburgh, NY. Project C-6075 OWRT, U.S. Department of the Interior, Washington, D.C.

Bormann, F.H., Likens, G.E. 1979. *Pattern and Process in a Forested Ecosystem.* Springer-Verlag, New York.

Bratton, S.P. 1974. The effect of the European wild boar (*Sus scrofa*) on the high-elevation vernal flora in Great Smokey Mountains National Park. *Bull. Torrey Botan. Club* 101:198–206.

Bratton, S.P. 1975. The effect of the European wild boar, *Sus scrofa* on gray beech forest in the Great Smokey Mountains. *Ecology* 56:1356–1366.

Chadbreck, R.H. 1958. Beaver-forest relationships in St. Tammany parish, Louisiana. *J. Wildlife Manag.* 22(2):179–183.

Champagne, G.C. 1971. The beaver in New York. *NYS. Conserv.* 26(1):18–21.

Connell, J.H., Slatyer, R.O. 1977. Mechanisms of succession in natural communities and their role in community stability and organization. *Am. Naturalist* 111(982):1119–1144.

Curtis, J.T. 1956. The modification of mid-latitude grasslands and forests by man, pp. 721–736. *In* W.L. Thomas (ed.), *Man's Role in Changing the Face of the Earth.* University of Chicago Press, Chicago, IL.

Daubenmire, R. 1968. *Plant Communities: A Textbook of Plant Synecology.* Harper and Row, New York.

Denslow, J.S. 1985. Disturbance-mediated coexistence of species, pp. 307–323. *In* S.T.A. Pickett and P.S. White (eds.), *The Ecology of Natural Disturbance and Patch Dynamics.* Academic Press, New York.

Dickinson, N.R. 1971a. Wetlands inventory—central Adirondacks section—Hamilton and Warren Counties. New York State Department of Environmental Conservation.

Dickinson, N.R. 1971b. Aerial photographs as an aid in beaver management. *NY Fish Game J.* 18(1):57–61.

Dix, R.L. 1959. The influence of grazing on the thin-soil prairies of Wisconsin. *Ecology* 40:36–49.

Egler, F.E. 1954. Vegetation science concepts: I. Floristic composition. A factor in old-field vegetation development. *Vegetatio* 4:412–417.

Forman, R.T.T., Boerner, R.E.J. 1981. Fire frequency and the Pine Barrens of New Jersey. *Bull. Torrey Botan. Club* 108:34–50.

Forman, R.T.T., Godron, M. 1981. Patches and structural components for a landscape ecology. *BioSci.* 31:733–740.

Forman, R.T.T., Godron, M. 1986. *Landscape Ecology.* Wiley & Sons, New York.

Gauch, H.G. 1977. ORDIFLEX—A flexible computer program for four ordination techniques: weighted averages, polar ordination, principal components analysis and reciprocal averaging Release B. Ecology and Systematics, Cornell University, Ithaca, NY.

Gauch, H.G. 1982. Multivariate analysis in community ecology. Cambridge Studies in Ecology, Cambridge University Press, MA.

Gill, J.D., Healy, W.M. 1974. Shrubs and vines for northeastern wildlife. Northeastern Forest Experiment Station, Forest Service, U.S. Department of Agriculture, USDA Forest Service General Technical Report # NE-9.

Golley, F.B., Ryszkowski, L., Sokur, J.T. 1975. The role of small mammals in temperate forests, grasslands and cultivated fields, pp. 223–240. *In Small Mammals: Their Productivity and Population Dynamics.* International Biological Programme, Vol 5. Cambridge University Press, MA.

Gruendling, G.K., Bogucki, D.J., Adams, K.B. 1985. Data collection for testing alternative hypotheses concerning increased acidification and fish population declines in complex Adirondack lake systems. Final report for Oak Ridge National Laboratory, Oak Ridge, TN.

Hall, A.M. 1971. Ecology of beaver and selection of prey by wolves in central Ontario. Master's thesis, University of Toronto, Ontario, Canada.

Harper, J.L. 1969. The role of predation in vegetational diversity. Brookhaven Symposium of Biology No. 22, pp. 48–62.

Heinselman, M.L. 1981. Fire and succession in the conifer forests of northern North America, pp. 374–405. *In* D.C. West, H.H. Shugart, and D.B. Botkin (eds.), *Forest Succession: Concepts and Application*. Springer-Verlag, New York.

Henry, J.D., Swan, J.M.A. 1974. Reconstructing forest history from live and dead plant material—an approach to the study of forest succession in southwest New Hampshire. *Ecology* 55:772–783.

Hodgdon, H.E. 1978. Social dynamics and behavior within an unexploited beaver (*Castor canadensis*) population. Doctoral dissertation, University of Massachusetts, Boston, MA.

Hodkinson, I.D. 1975a. Dry weight loss and chemical changes in vascular plant litter of terrestrial origin occurring in a beaver pond ecosystem. *J. Ecology* 63:131–142.

Hodkinson, I.D. 1975b. A community analysis of the benthic insect fauna of an abandoned beaver pond. *J. Animal Ecology* 44:535–553.

Hodkinson, I.D. 1975c. Energy flow and organic matter decomposition in an abandoned beaver pond ecosystem. *Oecologia* 21:131–139.

Horn, H.S. 1976. Succession, pp. 187–204. *In* R.M. May (ed.), *Theoretical Ecology: Principles and Applications*. Blackwell, London.

Howard, J.A. 1970. *Aerial Photo—Ecology*. Elsevier, New York.

Howard, R.J. 1982. Beaver habitat classification in Massachusetts. Master's thesis, University of Massachusetts, Amherst, MA.

Howard, R.J., Larson, J.S. 1985. A stream habitat classification system for beaver. *J. Wildlife Manage* 49(1):19–25.

Itow, S. 1963. Grassland vegetation in uplands of western Honshu, Japan.—II. Succession and grazing indicators. *Jpn. J. Botany* 18:133–167.

Ives, R.L. 1942. The beaver-meadow complex. *J. Geomorphol.* 5:191–203.

Jamieson, P. 1985. The Adirondack Park, pp. 11–23. *In* N. Farb (author), *The Adirondacks*. Rizzoli International Publishing, New York.

Johnson, C.E. 1927. The beaver in the Adirondacks: Its economics and natural history. *Roosevelt Wildlife Bull. NYS. Coll. Forestry, Syracuse, NY* 4(4):495–641.

Johnson, P.L. 1969. *Remote Sensing in Ecology*. University of Georgia Press, Athens, GA.

Kershaw, K.A. 1964. *Quantitative and Dynamic Ecology*. Elsevier, New York.

Kirby, W.R. 1976. Mapping wetlands on beaver flowages with 35-mm photography. *Can. Field Naturalist* 90(4):423–431.

Klecka, W.R. 1975. Discriminant analysis, pp. 434–467. *In* N.H. Nie, C.H. Hull, J.G. Jenkins, K. Steinbrenner, and D.H. Bent (eds.), *SPSS, Statistical Package for the Social Sciences*. McGraw-Hill, New York.

Knudson, G.J. 1962. Relationship of beaver to forests, trout and wildlife in Wisconsin. Technical Bulletin No. 25 Wisconsin Conservation Department, Madison, WI.

Map Information Unit. 1979. Inventory of aerial photography and other remotely sensed imagery of New York State (1968–1979). New York State Department of Transportation, Albany, NY.

McNaughton, S.J. 1979. Grassland-herbivore dynamics, pp. 46–81. *In* A.R.E. Sinclair and M. Norton-Griffiths (eds.), *Serengeti. Dynamics of an Ecosystem*. University of Chicago Press, Chicago, IL.

McNaughton, S.J. 1983. Serengeti grassland ecology: The role of composite environmental factors and contingency in community organization. *Ecol. Monogr.* 53:291–320.

Mooney, H.A., Godron, M. 1983. *Disturbance and Ecosystems: Components of Response*. Springer-Verlag, New York.

Naiman, R.J., Melillo, J.M. 1984. Nitrogen budget of a subarctic stream altered by beaver (*Castor canadensis*). *Oecologia* 62:150–155.

Naiman, R.J., McDowell, D.M., Farr, B.S. 1984. The influence of beaver (*Castor canadensis*) on the production dynamics of aquatic insects. *Internationale Vereinigung fur Theoretische und Angewandte Limnologie* 22:1801–1810.

Naveh, Z., Kinski, J. 1975. The effect of climate and management on species diversity of a Tabor oak-savannah pasture in Israel, pp. 284–296. *In Proceedings of the 6th Scientific Conference of the Israeli Ecological Society*. Tel Aviv, Israel.

Neff, D.J. 1957. Ecological effects of beaver habitat abandonment in the Colorado rockies. *J. Wildlife Manag.* 21(1):80–84.

Noble, I.R., Slatyer, R.O. 1980. The effect of disturbance on plant succession. *Proc. Ecol. Soc. Australia* 10:135–145.

Nyland, R.D., Remelle, K.E. 1975. Prism measures distance for plot boundaries. State University of New York, College of Environmental Science and Forestry AFRI Research Notes, Syracuse, NY.

Parsons, G.P., Brown, M.K. 1978. An assessment of aerial photograph interpretation for recognizing potential beaver colony sites. *NY. Fish Game J.* 25(2):175–177.

Peterman, R.M. 1978. The ecological role of mountain pine beetle in lodgepole pine forests and the insect as a management tool. *In* A.A. Berryman, R.W. Stark, and G.D. Amman (eds.), *Mountain Pine Beetle Management in Lodgepole Pine Forests*. University of Idaho Press, Moscow, ID.

Pickett, S.T.A., White, P.S. 1985. *The Ecology of Natural Disturbance and Patch Dynamics*. Academic Press, New York.

Platt, W.J. 1975. The colonization and formation of equilibrium plant species associations on badger disturbances in a tall-grass prairie. *Ecol. Monog.* 45:285–305.

Pritchard, G., Hall, H.A. 1971. An introduction to the biology of craneflies in a series of abandoned beaver ponds, with an account of the life cycle of *Tipula sacra* Alexander (Diptera: Tipulidae). *Can. J. Zoology* 49:467–482.

Radford, H.V. 1907. History of the Adirondack beaver, pp. 389–418. *In New York State Forestry, Fish and Game Commission Annual Report 1904, 1905, 1906*.

Remillard, M.M. 1984. Use of aerial photography to study beaver (*Castor canadensis* KUHL) impact on plant community succession in the Adirondacks of New York. Master's thesis, State University of New York, Plattsburgh, NY.

Risser, P.G., Karr, J.R., Forman, R.T. 1984. Landscape ecology, directions and approaches. Illinois Natural History Survey, Champaign, IL.

Romme, W.H. 1982. Fire and landscape diversity in subalpine forests of Yellowstone National Park. *Ecol. Monog.* 52:199–221.

Romme, W.H., Knight, D.H. 1981. Fire frequency and subalpine forest succession along a topographic gradient in Wyoming. *Ecology* 62:319–326.

Rue, L.L., III. 1964. *The World of the Beaver*. J.B. Lippincott, Philadelphia, PA.

Ruffner, J.A. 1980. *Climates of the States*. Gale Research Company, Detroit, MI., pp. 267–282, 530–555.

Seaton, E.T. 1928. *Lives of Game Animals*. C.T. Branford, Boston, MA.

Shadle, A.R. 1956. The American beaver. *Animal Kingdom* 58:98–104, 152–157, 181–185.

Sousa, W.P. 1984. The role of disturbance in natural communities. *Ann. Rev. Ecol. Systematics* 15:353–391.

Sprugel, D.G. 1976. Dynamic structure of wave-generated *Abies balsamea* forests in the northeastern United States. *J. Ecology* 64:889–911.

Terrie, P.G. 1985. *Forever Wild*. Temple University Press, Philadelphia, PA.

Turner, M.G. 1985. Ecological effects of multiple perturbations on a Georgia saltmarsh. Doctoral dissertation, University of Georgia, Athens, GA.

van der Valk, A.G., Davis, C.B. 1978. The role of seed banks in the vegetation dynamics of prairie glacial marshes. *Ecology* 59(2):322–335.

Vogl, R.J. 1974. Effects of fire on grasslands, pp. 139–194. *In* T.T. Kozlowski, and C.E. Ahlgren (eds.), *Fire and Ecosystems*. Academic Press, New York.

Vogl, R.J. 1981. The ecological factors that produce perturbation-dependent ecosystems, pp. 63–94. *In* D.C. West, H.H. Shugart, and D.B. Botkins (eds.), *Forest Concepts and Application*. Springer-Verlag, New York.

Weller, M.W. 1981. *Freshwater Marshes*. University of Minnesota Press, Minneapolis, MN.

Weller, M.W., Spatcher, C.E. 1965. Role of habitat in the distribution and abundance of marsh birds. Iowa State University Agriculture and Home Economics Experimental Station, Special Report Number 43.

Weller, M.W., Fredrickson, L.H. 1974. Avian ecology of a managed glacial marsh. *The Living Bird* 12:269–291.

White, P.S. 1979. Pattern, process, and natural disturbance in vegetation. *Botan. Rev.* 45:229–299.

Whittaker, R.H. 1977. *Animal Effects on Plant Species Diversity. Vegetation und Fauna*, Cramer, pp. 409–425.

Wilde, S.A., Youngberg, C.Y., Hovind, J.H. 1950. Changes in composition of ground water, soil fertility, and forest growth produced by the construction and removal of beaver dams. *J. Wildlife Manag.* 14:123–128.

7. Suppression of Natural Disturbance: Long-Term Ecological Change on the Outer Banks of North Carolina

William E. Odum, Thomas J. Smith, III, and Robert Dolan

7.1 Introduction

The artificial suppression of natural disturbance can often lead to a variety of side effects, ranging from alterations in geomorphology to significant changes in ecological productivity and community structure and structural properties at the landscape level of organization. As White (1979) has noted, these alterations are frequently unanticipated and detrimental. A few familiar examples include: (1) forest fire suppression leading to buildup of fire fuels and unnaturally severe fires (Mutch 1970; Niering 1981); (2) suppression of the annual floods of the Colorado River resulting in erosion and loss of beaches (Dolan et al. 1974); and (3) suppression of periodic flooding of swamp forests leading to reduced primary productivity (Conner and Day 1976).

Because of two contrasting dune management practices, the Outer Banks of North Carolina present an unusual opportunity to investigate the effects of suppression of one of the major coastal natural disturbances, storm-induced flooding and oceanic overwash and associated salt spray. One section of the landscape closely resembles the original, natural condition and has never been altered significantly or managed. On another section a stabilized dune system was built, planted with vegetation, and managed for approximately 30 years before management was terminated in 1972.

In effect, the stabilized dune system appears to be an introduced landscape element acting to retard the spread of natural disturbance. To better understand this phenomenon, we began a comparative study in 1971 of the natural and stabilized dune systems. Our objectives were:

1. To document how the landscape was altered by the creation of an unnatural corridor.
2. To investigate how suppression of the natural disturbance regime affected the geomorphology, vegetation dynamics, animal diversity, and landscape heterogeneity.
3. To monitor the long-term effects of the gradual decline of the stabilized dune system after abandonment of most management practices in 1972.

We investigated these questions by surveying and sampling vegetation and animals on transects across both dune systems in 1971, while the management program was still in effect and again in 1978 and 1984, six and 12 years after management ceased.

7.2   Creation of the Dune Corridor

The Outer Banks of North Carolina are a series of narrow, low-lying barrier islands characterized by frequent, storm-induced changes, including beach and marsh erosion and rebuilding, dune formation and destruction, and inlet openings and closings (Godfrey and Godfrey 1976). During extreme storms the combination of high tides and great wave energy is capable of transporting beach sand and broken shell completely across the islands; this process of "oceanic overwash" (discussed by Godfrey, 1970) made it impossible to construct a permanent road system on the Outer Banks before the 1930s.

To provide protection for a permanent highway, an artificial dune system was designed to lie between the beach and the proposed highway. Between 1936 and 1940 the Civilian Conservation Corps, under the direction of the National Park Service, erected approximately 1 million m of sand fencing to create a nearly continuous barrier dune along the upper portion of the Outer Banks (Stratton and Hollowell 1940).

To prevent the barrier dune from blowing away, the entire structure was stabilized with a mixture of trees, shrubs, and grasses. The most successful vegetation proved to be American beach grass, *Ammophila breviligulata*, and to a lesser extent, sea oats, *Uniola paniculata*. With proper management American beach grass, although planted south of its original range, has proven to be well suited for sand stabilization; it lends itself to nursery propagation, can be transplanted with almost perfect survival, grows rapidly after transplanting, and is capable of trapping and growing up through large quantities of sand (Woodhouse and Hanes 1966). Sea oats take much longer to become established (three to five years), but often replace American beach grass on the dune crest, even in carefully cultivated plots (Cooper 1972).

The barrier dune system was further augmented in the late 1950s by the National Park Service; included at this time was an extension of the system to encompass Ocracoke Island. The final result was an almost continuous dune

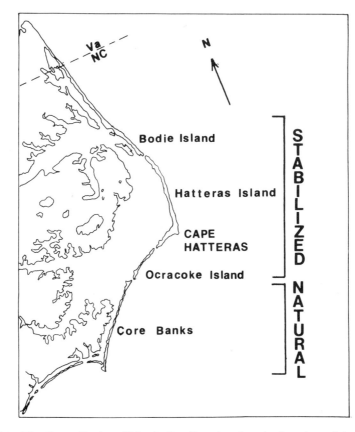

Figure 7.1. The Outer Banks of North Carolina showing the location of the stablized islands (part of Bodie Island, Hatteras Island including Pea Island, and Ocracoke Island) and the unaltered, natural Core Banks.

stretching 110 km from south Nags Head to the southern tip of Ocracoke Island (Fig. 7.1). Maintenance of this system involved periodic fertilization, spot replanting, and dune rebuilding to repair storm damage. Maintenance costs were high, totaling more than $20 million between 1950 and 1970 (Behn and Clark 1979).

This continual expense caused both Congress and the National Park Service to re-examine the artificial stabilization program between 1969 and 1972. High maintenance costs combined with scientific arguments against the long-term benefits of the program (reviewed by Dolan et al. 1973) resulted in the cessation of dune management by the National Park Service on the Outer Banks after 1972 (Behn and Clark 1979).

The region south of Ocracoke Island, from Portsmouth Island to Cape Lookout (Fig. 7.1), known collectively as the Core Banks, remains relatively undeveloped and unaltered. Because the Core Banks do not have a stabilized barrier dune, they are nearly in a natural state. They are characterized by broad beaches; scattered, low, and irregular dunes; and frequent instances of oceanic overwash extending as far as the marshes on the soundside shore of the islands. Thus, for comparative purposes the banks can be divided into two sections: (1) a northern portion which has been altered through the dune stabilization program and now exists in an altered but largely unmaintained state, and (2) a southern stretch which closely approximates the natural condition.

7.3 Research Approach

In 1971 we initiated a long-term sampling program at three sites on the Outer Banks: (1) the Core Banks site, a relatively undisturbed dune system approximately 8 km north of the Cape Lookout lighthouse; (2) the Pea Island site, a single stabilized dune ridge built about 1938 and located 4 km south of the Oregon Inlet Bridge; and (3) the Ocracoke Island site, a double dune system built about 1958 and adjacent to milepost 69.5 on the Ocracoke highway. At each site we surveyed permanent transects from the beach berm across the dune tract and through the flats behind the dunes to near the center of each island. Transect lengths varied from 148 to 278 m.

Quadruplicate samples were taken along each transect at 2-m intervals for percent cover (with a movable sighting device), litter, and standing biomass. Shrubs were not sampled, but their areal extent was estimated using time series aerial photography. Small mammals were trapped on a grid, and bird density was estimated with the method of Eberhardt (1978). The transects were sampled intensively in July and August of 1971 and 1978 and revisited for qualitative observations in July 1984 and January 1986. Data from the 1971 Core Banks site were not used in subsequent analyses because the transect was relocated between 1971 and 1978. Statistical comparisons were made with the Statistical Package for the Social Sciences Analysis of Variance (SPSS ANOVA) program (Nie et al. 1975).

7.4 Research Findings

7.4.1 Altered and Unaltered Landscapes

7.4.1.1 Geomorphology

A cross-sectional comparison of the three sites (Fig. 7.2) shows marked geomorphological differences. The unaltered site on the Core Banks is characterized by a broad beach and small isolated dunes less than 1 m in height. These dunes are roughly circular or oval in shape and never present a continuous barrier to oceanic overwash. In marked contrast, the stabilized dunes at Ocracoke and Pea Island form continuous ridges parallel to the ocean and many kilometers in length. The Ocracoke dune system was constructed in 1958 and presently has an elevation of 3 to 5 m. The Pea Island system was constructed during

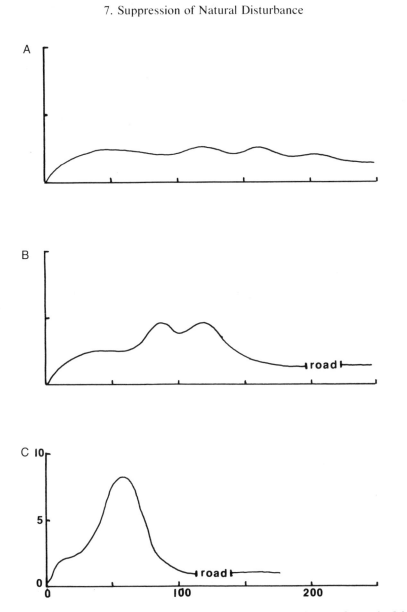

Figure 7.2. Transects across the three study sites: (A) Core Banks, (B) Ocracoke Island, (C) Pea Island (northern end of Hatteras Island). Elevations and distances are in meters; scales are identical on all three transects.

1936 to 1940. With more time for sand accumulation, by the mid 1970s it had reached a height of approximately 10 m. As discussed elsewhere in this paper, this trend has been reversed somewhat by the cessation of active management, with a gradual decrease in dune height at Pea Island during the 1980s.

With the exception of small, isolated dunes, the beach at the Core Banks site averages about 100 m in width. At Ocracoke the beach is only 60 m wide, whereas at Pea Island it has shrunk to 20 m or less. These observations support the hypothesis of Dolan et al. (1973), who suggested a gradual reduction of beach width and increased wave attack on the dune face as a result of rising sea level and lack of room for inland beach migration caused by the dune stabilization program. By the late 1970s the stabilized dune face at both Pea Island and Ocracoke had been extensively eroded by wave action, a condition which did not exist in 1971.

7.4.1.2   Plant Community Dynamics

Detailed data documenting differences in plant community characteristics between the three sites and before and after management are given in Odum et al. (in preparation). The following is a short summary of that data.

Plant biomass varied significantly ($P = .01$) between the three sites. At Pea Island in 1971 there was a relatively high biomass (mean, 880 ± 117 dry g/m^2 on the dune backs), extending completely across the dune system and onto the flats behind the dunes. At Ocracoke this was considerably lower (mean, 225 ± 49 g/m^2 on the dune backs), and at the Core Banks it was even lower (97 ± 16 g/m^2 on the dune backs).

By 1978, six years after the cessation of active management (fertilization and replanting of *Ammophila breviligulata*), average biomass estimates had declined by 50% or more at Pea Island but had not changed significantly at the Ocracoke Island site, which had been managed less intensively than Pea Island.

Plant species diversity (species number, H', dominance concentration) did not vary significantly between the dune communities at the three sites, although diversity was significantly higher on the flats behind the dunes at the natural Core Banks site (for example H' was 1.38 at the Pea Island flats and 2.28 at Core Banks). In 1978, six years after the cessation of intensive management, species diversity had risen significantly at all sections of the dunes at Pea Island, but not on the dune flats at Pea Island or at Ocracoke.

Total vegetative cover exceeded 85% at both managed sites (Pea Island and Ocracoke) in 1971 but was less than 40% on the Core Banks dunes. By 1978 the vegetative cover at Pea Island had dropped to 60 to 70% on the dune fronts and tops but remained above 80% on the protected dune backs and flats behind the dunes. Significant changes in vegetative cover did not occur at Ocracoke. Plant litter was not sampled in 1971 but was much higher at Pea Island in 1978 (537 ± 80 g/m^2 on dune backs) compared with Ocracoke (31 ± 9 g/m^2 on dune backs) and the Core Banks (22 ± 8 g/m^2 on dune backs).

Many plants species occurred much closer to the high water mark on the managed transects than on the natural transect at the Core Banks. For example, the shrub *Myrica cerifera* was as close as 59 m from the mean high tide mark at Pea Island, but was 220 m away at Ocracoke and 249 m at the Core Banks. The grass, *Spartina patens*, was only 61 m from water at Pea Island, 105 m at Ocracoke, and 104 m at the Core Banks. Similar patterns were found for almost all other species of plants.

Time series aerial photography covering the period 1950 to 1980 shows expansion of the shrub communities (*Myrica cerifera, Ilex vomitoria, Baccharis halimifolia,* and *Juniperus virginiana*) on the flats behind the stabilized dunes and even onto the dune backs (also noted by Shroeder et al. 1976; Bellis 1980). Expansion of individual shrubs and patches of shrubs has been highly irregular and apparently determined by more factors (e.g., soil type and soil moisture) than simply protection from salt spray and flooding afforded by the stabilized dune line.

7.4.1.3 Causes of Changes in the Dune Plant Community

The altered dune plant community (increased biomass, vegetative cover and plant litter, decreased diversity, expansion of shrub patches, and closer approach of many plant species to the active beach) on the stabilized dunes appears to be related to a combination of factors: (1) increased protection from flooding afforded by the unnaturally elevated artificial dune; (2) increased protection from salt spray, particularly on the back slope of the stabilized dune (documented by van der Valk 1974a); (3) the systematic planting and replanting of *Ammophila;* and (4) artificial fertilization.

There are, of course, other possible factors such as alterations in soil properties and subsoil hydrology caused by artificial dune construction and interference with nutrient inputs via salt spray behind the stabilized dune (van der Valk 1974b). The effects of all of these factors have been further modified by the interspecific interactions inherent in any dune plant communities (i.e., Silander and Antonovics 1982). Because all of these factors interact in a synergistic fashion, it is difficult to isolate a single "master" variable.

Our estimates of plant species diversity are similar to those found by van der Valk (1975) from stabilized dune communities in the same geographic area. The tendency for lowered species diversity to result from planting and fertilization is a general finding from a variety of ecosystems including old fields (Mellinger and McNaughton 1975; Bakelaar and Odum 1978), lawns (Falk 1976), and postfire forests (Bock et al. 1978).

Significant differences in plant community structure between the older managed site at Pea Island and the more recent site at Ocracoke can be traced to: (1) the greater age (and height) of the Pea Island dune ridge; (2) differences in the methods used to build and plant dunes (single versus double dunes and different combinations of plant species; van der Valk 1975); (3) differences in exposure and microclimate (Pea Island faces east and is a little cooler in summer than Ocracoke, which faces south); (4) the failure of some of the *Ammophila* plantings at Ocracoke; and (5) differences in plant interspecific interactions at the two sites.

In reference to point 4, the failure of introduced *Ammophila* to persist in a region well south of its normal range (Cooper 1972) is not surprising, particularly because we noted infestations of the scale insect *Eriococcus carolinae* (Campbell and Fuzy, 1972) at Ocracoke as early as 1971. In addition, *Uniola* appears to have a competitive advantage on top of the Pea Island dune (also noted by van der Valk 1975; Silander and Antonovics 1982); throughout the Ocracoke dune system; and other hot, dry sites (Woodhouse et al. 1968), possibly because it is a C_4 photosynthetic pathway species, whereas *Ammophila* is a C_3 (Mayne, personal communication).

In reference to point 5, the results of Silander and Antonovics (1982), who used single and multiple species removal experiments, emphasize the importance of interspecific interactions in controlling dune community structure. Their results further suggest that interspecific interactions may be equally or more important than abiotic factors (flooding, salt spray, etc.) in controlling dune plant community structure in relatively benign locations such as the lower portions of the dune back slope.

7.4.1.4   Related Changes in the Animal Community

Forman and Godron (1986) have hypothesized that increasing landscape heterogeneity should increase animal species diversity due to increased incidence of "edge" species, but that rare, interior species found in homogeneous environments may be lost. Our bird data (Table 7.1) tends to support this idea in a very preliminary way as species diversity is higher at the more heterogeneous sites at Ocracoke and Core Banks.

Small mammals were not a good test of the heterogeneity-diversity hypothesis because only three species were captured at all of the sites combined (*Mus musculus, Oryzomys palustris*, and *Peromyscus leucopus*), and only *Mus* was common in the dune tract. It is interesting to note that densities of *Mus* were six to eight times higher at the stabilized dune sites than at the natural dune site, a finding similar to that of Dueser et al. (in preparation).

7.4.2   Cessation of Dune Management

Six years without dune management at Pea Island and Ocracoke resulted in only slight geomorphological change. However, significant ecological changes had occurred, suggesting an intermediate state between the previous managed condition and the natural condition as represented by the Core Banks. These changes included: (1) decreases in plant biomass to an intermediate level; (2) replacement of *Ammophila* by native species, such as *Spartina patens* and *Uniola;* (3) increased species richness, increased H', and decreased dominance concentration; (4) a decreased, but intermediate value of vegetation percent cover; (5) the continued presence of a significant litter layer; and (6) the persistence of many plant species unusually close to the water's edge.

Qualitative observations made in August 1984 and January 1986, 12 and 14

Table 7.1. Densities of Birds (#/hectare) on the Three Transects Estimated with the Method of Eberhardt (1978)

	Pea Island	Ocracoke Island	Core Banks
Red-winged	Not observed		13.6
blackbird		7.2	
Barn swallow	Not observed	7.4	11.0
Mourning dove	Not observed	0.9	0.6
Seaside sparrow	6.5	Not observed	Not observed

years after the cessation of management, suggest a gradual change in the geomorphology of the stabilized dunes (e.g., eroded dune faces, lowered dune height, occasional "blow-outs" or bare areas). Moreover, the shift toward an intermediate state in the plant community structure is firmly entrenched (i.e., the first four trends mentioned in the preceding paragraph are continuing). By 1986 shrub patches behind the dune ridge showed extensive damage from salt spray and flooding associated with the passage of hurricane Gloria in 1985.

We suspect that the dune vegetation on formerly managed dune tracts of the Outer Banks will continue in this trend of gradual change until physical processes have reduced the artificial dune line to the original state of small, low scattered dunes lying behind a broad, open beach. Occasional intervention to rebuild openings in the dune line without subsequent intensive management (i.e., fertilization, etc.) could prolong this intermediate state. In fact, this may be the strategy adopted in the future by the National Park Service.

7.5 Effects of Disturbance on Landscape Heterogeneity

The natural, unaltered Hatteras barrier islands (e.g., the Core Banks) have a landscape structure composed of a matrix of bare sand upon which three elements are superimposed: (1) small patches of dune vegetation, (2) small patches of shrubs and small trees, and (3) two bands, one of shrubs and one of salt marsh vegetation on the sound side of the island (Fig. 7.3A). This produces a landscape on the seaward side of the island that is relatively "fine-grained." In other words, the sizes of these vegetation patches tend to be very small, ranging from a meter or less to a few tens of meters.

On the stabilized banks (e.g., Pea Island) the situation is considerably different. The construction of the stabilized sand dune system along with subsequent management practices has created two longitudinal bands or corridors toward the front of the island (Fig. 7.3B). One corridor is composed of the stabilized dune with its covering of predominantly *Ammophila*. The second corridor consists of the shrub zone, which has developed behind the protective influence of the stabilized dune and is made up of unnaturally dense patches of shrubs close to the beach. Both of these bands can be relatively broad (tens to hundreds of meters) and extend for tens of kilometers. The result is a coarser grained landscape pattern on the ocean side of the island than originally existed before the stabilization program. Since the termination of management in 1972, these corridors are gradually breaking up, producing a finer grained landscape (Fig. 7.3C).

Forman and Godron (1986) have hypothesized that increasing disturbance can lead to increasing landscape heterogeneity, and that very high levels of disturbance can result in either high or low heterogeneity. Our data from the Outer Banks tend to at least partially support this hypothesis. For example, on the natural Core Banks, the highest levels of landscape heterogeneity appear to occur in the center and toward the rear of the island (Fig. 7.3A) where disturbance from storms, spray, and overwash is of intermediate intensity. Heterogeneity decreases toward the beach and becomes extremely low (highly ho-

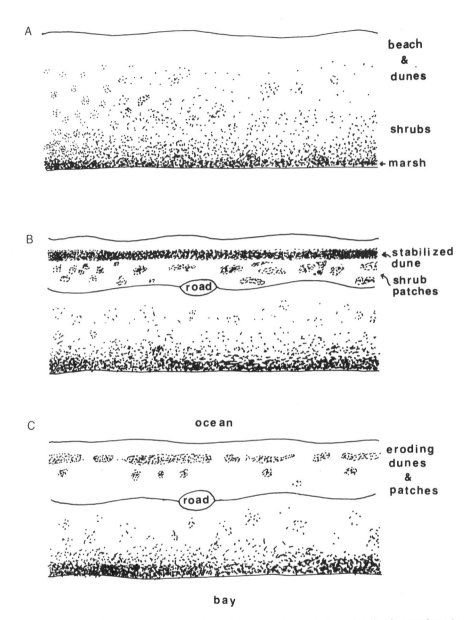

Figure 7.3. A schematic representation of dune and vegetation distribution under (A) natural conditions as found on the Core Banks, (B) stabilized conditions such as existed at Pea Island and Ocracoke Island in 1971, and (C) partially recovering conditions found in 1986 at the formerly stabilized sites. Features as shown are generalized from many aerial photographs and do not represent conditions at specific sites.

mogeneous) on the bare beach with its daily and monthly wave and storm perturbation.

On the stabilized banks (Fig. 7.3B), the situation is somewhat different. Construction of the stabilized dune, despite reducing levels of perturbation, has increased landscape heterogeneity toward the front of the island by creating a dune grass corridor, a shrub corridor behind it, and associated patches of shrubs, grasses, and small trees. The middle and soundside portions of the island appear to remain at the same approximate level of landscape heterogeneity. Probably, given a long enough period without management, the forefront of the island would decrease in landscape heterogeneity as the complex pattern of corridors and patches gradually reverts to scattered dunes and bare sand.

7.6   Resistance and Resilience

Webster et al. (1975) define "resistance" as the ability to absorb disturbance without changing (i.e., to "resist" the disturbance). "Resilience" is defined as the ability to return to nominal state after disturbance.

An ecosystem that is very resilient but has low resistance will tend to fluctuate greatly but persist for an extended time. Conversely, an ecosystem with high resistance but low resilience will rarely fluctuate, but may not persist if severely stressed. In other words, ecosystems of this type have a high risk of permanent alteration.

Although using different terminology for the same concepts, Holling (1973) points out that ecosystem management should be based on "resilience" rather than "stability" (i.e., "resistance" in Webster's terminology). In other words, we should encourage and devise ecosystems that can absorb and accommodate stress and continue to persist.

In the context of the preceding discussion, the natural barrier island dune system has low "resistance" but high "resilience." Frequent storm-induced fluctuations occur in both geomorphology and plant community structure, but the ecosystem almost always absorbs the changes, recovers, and persists. The managed dunes at Pea Island and Ocracoke, however, have high "resistance" but low "resilience." Ecological and geomorphological fluctuations rarely occur, but when they do (i.e., breaching and destruction of the dune by storm overwash), the ecosystem is permanently altered and fails to persist in its original state. The only way for the managed dune ecosystem to return to its former state is for humans to intervene and, at considerable expense, re-establish the artificial state.

7.7   Conclusions

The creation of an unnatural corridor in the form of a continuous, heavily vegetated, stabilized dune system has had a number of ecological consequences. This is because the corridor acted as a barrier to natural disturbances such as flooding, oceanic overwash, and salt spray, which normally determine the barrier island landscape.

Comparisons made one year before and six years after cessation of dune management indicate that suppression of natural disturbance leads to (1) moderately increased plant biomass, (2) moderately increased percent cover, (3) a significantly increased litter layer, (4) occurrence of many species much closer to the water line than expected, (5) the proliferation of a dense shrub corridor behind the stabilized dune line, (6) an increase in landscape heterogeneity toward the front of the island, and (7) a coarser grained landscape structure toward the front of the island.

Management practices on the stabilized dunes, which included the planting of American beach grass, *Ammophila breviligulata*, periodic fertilization, and dune reconstruction after damaging storms, caused (1) a decline in plant species richness and H' diversity, (2) an increase in plant dominance concentration, (3) a greatly increased plant biomass (and presumably annual productivity), (4) an almost complete plant coverage of the dune surface, and (5) contributed to the increase in landscape heterogeneity and the coarser grained landscape structure on the front half of the islands. Cessation of management practices in 1972 appears to have caused a reversal in the trends of all of these variables.

The stabilized, managed dune tract appears to be a system with high resistance but low resilience to stress in the form of storm-induced flooding, overwash, and salt spray. In contrast, the natural dune tract has low resistance but high resilience to the same stresses.

Acknowledgments

Paul Godfrey of the University of Massachusetts provided invaluable assistance in helping with the design of our sampling methodology in 1971. Much of this research was supported by National Park Service contract No. PX-0001-8—64 to the senior author.

References

Bakelaar, R.G., Odum, E.P. 1978. Community and population level responses to fertilization in an old-field ecosystem. *Ecology* 59:660–665.

Behn, R.D., Clark, M.A. 1979. The termination of beach erosion control at Cape Hatteras. *Public Policy* 27:99–127.

Bellis, V. 1980. The vegetative cover on the barrier islands of North Carolina. *Veroff. Geobot. Inst. ETH. Stiftung. Rubel, Zurich* 69:121–144.

Bock, J.H., Raphael, M., Bock, C.E. 1978. A comparison of planting and natural succession after a forest fire in the northern Sierra Nevada. *J. Appl. Ecol.* 15:597–602.

Campbell, W.V., Fuzy, E.A. 1972. Survey of the scale insect effect on American beach grass. *Shore Beach* April:18–19.

Conner, W.H., Day, J.W. 1976. Productivity and composition of a bald cypress—water tupelo site and a bottomland hardwood site in a Louisiana swamp. *Am. J. Bot.* 63:1354–1364.

Cooper, A.W. 1972. The ecology and management of coastal dunes and dredge spoil in North Carolina. *In The Ecology of the Land*. Research Monograph No. 5. VPI, Blacksburg, VA.

Dolan, R., Godfrey, P.J., Odum, W.E. 1973. Man's impact on the barrier islands of North Carolina. *Am. Scientist* 61:152–162.

Dolan, R., Howard, A.G., Gallenson, A. 1974. Man's impact on the Colorado River in the Grand Canyon. *Am. Sci.* 62:392–401.

Eberhardt, L.L. 1978. Transect methods for population studies. *J. Wildl. Mgmt.* 42:1–31.

Falk, J.H. 1976. Energetics of a suburban lawn ecosystem. *Ecology* 57:141–150.

Forman, R.T.T., Godron, M. 1986. *Landscape Ecology.* Wiley & Sons, NY.

Godfrey, P.J. 1970. Oceanic overwash and its ecological implications on the Outer Banks of North Carolina. U.S. Dept. Interior, National Park Service Reprint, Office of Chief Scientists.

Godfrey, P.J., Godfrey, M.M. 1973. Comparison of geological and geomorphic interactions between altered and unaltered barrier island systems in North Carolina. *In* D.R. Coates (ed.), *Coastal Geomorphology,* State University of NY.

Godfrey, P.J., Godfrey, M.M. 1976. Barrier island ecology of Cape Lookout National Seashore and vicinity, NC. Nat. Park Serv. Monograph Series No. 9.

Holling, C.S. 1973. Resilience and stability of ecological systems. *Ann. Rev. Ecol. Sys.* 4:1–23.

Mellinger, M.V., McNaughton, S.J. 1975. Structure and function of successional vascular plant communities in central New York. *Ecol. Monogr.* 45:161–182.

Mutch, R.W. 1970. Wildland fires and ecosystems—a hypothesis. *Ecology* 51:1046–1051.

Nie, N.H., Hull, C.H., Jenkins, J.G., Steinbrenner, K., Bent, D.H. 1975. *Statistical Package for the Social Sciences.* McGraw-Hill, New York.

Niering, W.A. 1981. The role of fire management in altering ecosystems. *In Fire Regimes and Ecosystem Properties.* U.S. Dept. Ag. Gen. Tech. Rept. WO-26, pp. 489–510.

Odum, W.E., Smith, III, T.J., Dolan, R. 1987. A comparison of plant communities on stabilized and natural dunes in North Carolina. *J. Coastal Res.* in press.

Schroeder, P.M., Dolan, R., Hayden, B.P. 1976. Vegetation changes associated with barrier-dune construction on the Outer Banks of North Carolina. *En. Mg.* 1:105–114.

Silander, J.A. 1979. Microevolution and clone structure in *Spartina patens. Science* 203:658–660.

Silander, J.A., Antonovics, J. 1982. Analysis of interspecific interactions in a coastal plant community—a perturbation approach. *Nature* 298:557–560.

Stratton, A.C., Hollowell, J.R. 1940. Sand fixation and beach erosion control. U.S.D.I. National Park Service. Office of Chief Scientist.

Webster, J.R., Waide, J.B., Patten, B.C. 1975. Nutrient cycling and the stability of ecosystems, pp. 1–27. *In* F.G. Howell, J.B. Gentry, and M.H. Smith (eds.), *Mineral Cycling in Southeastern Ecosystems.* ERDA Symposium Series, CONF-740513.

White, P.S. 1979. Pattern, process and natural disturbance in vegetation. *Botan. Rev.* 45:229–298.

Woodhouse, W.W., Hanes, R.E. 1966. Dune stabilization with vegetation on the outer banks of North Carolina. Soils Information Series 8, Dept. of Soil Science, N.C. State Univ., Raleigh, NC.

Woodhouse, W.W., Seneca, E.D., Cooper, A.W. 1968. Use of sea oats for dune stabilization in the southeast. *Shore Beach* 36:15–22.

van der Valk, A.G. 1974a. Environmental factors controlling the distribution of forbes on coastal foredunes in Cape Hatteras National Seashore. *Can. J. Bot.* 52:1057–1073.

van der Valk, A.G. 1974b. Mineral cycling in coastal foredune plant communities in Cape Hatteras National Seashore. *Ecology* 55:1349–1358.

van der Valk, A.G. 1975. The floristic composition and structure of foredune plant communities of Cape Hatteras National Seashore. *Ches. Sci.* 16:115–126.

van der Valk, A.G. 1977. The macroclimate and microclimate of coastal foredune grasslands in Cape Hatteras National Seashore. *Int. J. Biometeorol.* 21:227–237.

8. Vegetation Dynamics in a Southern Wisconsin Agricultural Landscape*

D.M. Sharpe, G.R. Guntenspergen, C.P. Dunn, L.A. Leitner, and F. Stearns

8.1 Introduction

The vegetation of southern Wisconsin was a complex mosaic before settlement began in the 1830s (Curtis 1959; Davis 1977; Finley 1976). The pattern of prairie, savanna, marshes, and forests responded to a number of site conditions and disturbance regimes that were themselves under the influence of climatic fluctuations (Kline and Cottam 1979) so that the mosaic was, itself, dynamic. Thus, the pattern of vegetation that the agricultural settlers encountered in the 1800s was the result of a unique history of interaction between vegetation, environment, and disturbance regimes. The settlers then commenced to remove a large fraction of the extant natural vegetation, and to fragment and subject the remainder to a markedly different disturbance regime. We expect that the forest remnants show the impact of both presettlement and postsettlement disturbance regimes in this landscape.

Unlike most of southern Wisconsin, a good record of the history of this deforestation is available for southwestern Green County. Shriner and Copeland (1904) mapped remnant forest tracts greater than or equal to 4 hectares in size for 1882 and 1902. Curtis (1956) extended this record to 1950. We have brought

*This research was supported by the National Science Foundation, Ecology Program, through grant no. DEB-8214792 to University of Wisconsin–Milwaukee, and grant no. DEB-8214702 to Southern Illinois University at Carbondale.

the record up to date for Cadiz Township. These records show that deforestation was rapid, and that it resulted in isolated woodlots that are generally small and uniformly distributed. The surrounding land is largely agricultural, with the pastures and tilled fields characteristic of dairying. Woodlots are managed as integral parts of farms and have been logged and/or grazed to varying degrees. They are typical of woodlots in agricultural regions of the Midwest.

Previous studies of the upland forests of southern Wisconsin form the basis for an ecological interpretation of present day patterns. These studies (Curtis and McIntosh 1951; Bray and Curtis 1957; Auclair and Cottam 1971; Peet and Loucks 1977) discuss change in vegetation as a result of changes in the disturbance regime, such as (1) an altered fire regime, (2) outbreaks of oak wilt, and (3) logging and grazing. Peet and Loucks (1977) also concluded that site factors could explain much of the variation in species stand composition from one stand to another. For the most part, these studies were in homogeneous oak-dominated woodlots or portions of woodlots (however, see Rogers 1959; Harper 1963) and have not addressed questions relating to: (1) species composition and dynamics of highly disturbed woodlots, which account for a majority of the forest area of agricultural regions and may therefore be more important to regional vegetation dynamics than undisturbed stands; and (2) the implications of the spatial distribution of stands that result from regional forest fragmentation.

In this chapter, we discuss ways in which the presettlement vegetation of three townships in southern Wisconsin have been transformed by agricultural development. We deal with three questions: (1) What was the distribution of presettlement vegetation, and of major tree species, and how did this relate to the distribution of sites? (2) Have particular site types been deforested selectively? (3) How does the current distribution of tree species compare with their presettlement distribution? Finally, we discuss how our results relate to a new interpretation of forest dynamics in agricultural landscapes, and research issues that they raise.

8.2   Methods

8.2.1   Study Area

The study area is comprised of three townships in south central Wisconsin (see Fig. 8.1, inset, for general location; Fig. 8.2 for three townships). Townships are the major unit of the rectangular land survey that was implemented in the central and western United States. A township is approximately 6 miles square (10 km × 10 km). The three townships are in the first tier of townships north of the Illinois–Wisconsin state line. Wayne Township is in southeast Lafayette County. Cadiz Township, to the east, is in southwest Green County, and Clarno Township is to the east of Cadiz Township. Cadiz Township is the focus of our research, and Wayne and Clarno Townships have been studied to place Cadiz Township in a broader context.

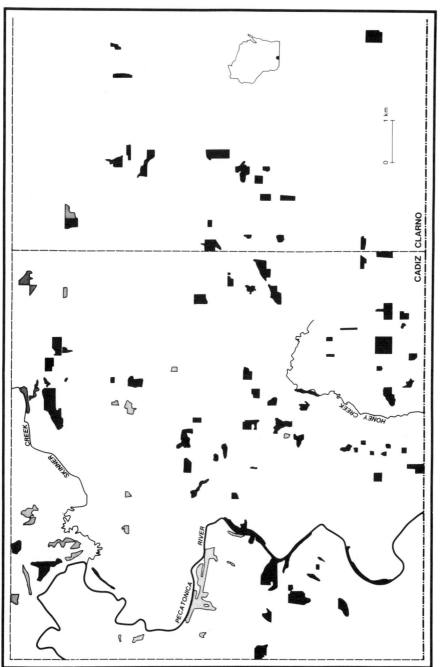

Figure 8.1. Location of forest tracts in the field survey, Cadiz and western Clarno Townships, Wisconsin. Forest tracts shown in black have been field surveyed; those with shaded pattern have not been surveyed.

PRESETTLEMENT VEGETATION
(1833)

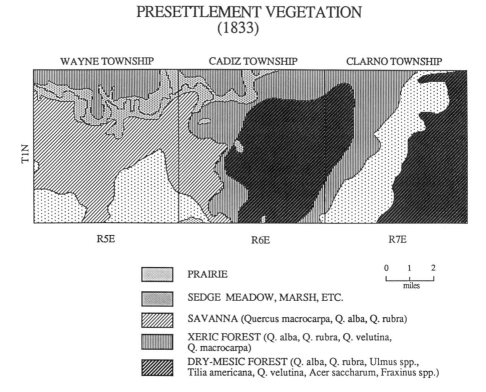

Figure 8.2. Presettlement vegetation in Wayne, Cadiz, and Clarno Townships, south central Wisconsin.

This study is based on three types of data: (1) witness tree records for Wayne, Cadiz, and Clarno Townships found in the General Land Office (GLO) land surveyors notes; (2) a spatial data base for Cadiz Township that includes environmental data and a time series of land use; and (3) a field survey of the forest tracts in Cadiz Township and western Clarno Township.

8.2.2 Presettlement Vegetation

Witness tree data were obtained from GLO land surveyors' notes for Cadiz Township, and for Wayne Township to the west and Clarno Township to the east. Witness trees were plotted on township maps by species, diameter, and distance to section corners. References to vegetation along survey lines were used to supplement witness tree data. Tree density and basal area at each point were determined using a modification of the quarter method (Anderson and Anderson 1975). The presettlement vegetation of the three townships was classified into prairie; sedge meadow and marsh; savanna; xeric forest; and dry mesic forest on the township maps. Finley's (1976) map of original vegetation was consulted while drawing the presettlement vegetation map (Fig. 8.2). The result is a generalized view of presettlement vegetation which generally does not take into consideration local patchiness that is not related to major topographic features.

Witness tree data were summarized separately for the portion of Cadiz Township east and west of the Pecatonica River. Importance value of each species was derived from these data as the average of relative density and relative basal area.

8.2.3 Spatial Data Base

The spatial data base for Cadiz Township includes land use and environmental data for each 1.0 hectare cell (approximately 10,000 cells) in the township. The distribution of presettlement vegetation is based on the township level maps that Finley (1976) used for his map of original vegetation of Wisconsin. Locations of remnant forest in 1882 and 1902 are based on maps by Shriner and Copeland (1904). Forest locations in 1937, 1956, 1963, and 1978 are based on aerial photographs and topographic maps. Soil type of each cell was obtained from the soil survey for Green County (Glocker 1974). A number of soil and site characteristics (e.g., soil texture, depth of A and B horizons, depth to bedrock and groundwater, available soil moisture, drainage, and slope) were derived from the description of each soil type in the soil survey. Surface hydrologic features (streams, rivers, lakes) were obtained from topographic maps.

The Cadiz Township spatial data base was obtained by overlaying the data source (maps, aerial photographs) with a mylar sheet marked with grid cells equal to 1.0 hectare at a scale of 1:24,000. Where necessary, the scales of the maps and aerial photographs were adjusted photographically to this scale. The grid system for the spatial data base was used as the standard for locating all of the forest tracts in Cadiz Township. The grid system was extended eastward into Clarno Township to locate forest tracts on the same grid system as those in Cadiz Township, but Clarno Township is not included in the spatial data base, per se.

The spatial data base has been subjected to a variety of analyses. IMGRID (Sinton 1977) and MAP (Tomlin 1983) were used to determine size and isolation of forest tracts. Environmental variables were used in cluster analysis (Radloff and Betters 1978) using the FASTCLUS procedure (SAS 1985) to assign each cell to a site type, which could then be correlated to its vegetation or land use. The variables used in the cluster analysis are shown in Table 8.1.

Ten clusters, or site types, were derived. The choice of the number of clusters is arbitrary, but we selected this number on the basis of the cubic clustering criterion described in SAS (1985). Site types were characterized and labeled with reference to the dominant soils in each. For example, one site type was dominated by Fayette and Palsgrove silt loams, which are characteristic of broad ridges and upper slopes. This site type was labeled "broad ridges."

Site types were associated with data in the spatial data base on presettlement vegetation to determine the extent of site selectivity for the major vegetation types. The chi-square test was used to assess whether all site types had the

Table 8.1. Variables for Cluster Analysis for Site Classification in Cadiz Township, Wisconsin

	Range	
Soil related		
Soil texture	1 = gravel	18 = clay
Soil depth	inches	
Available soil moisture	inches	
Soil drainage class	1 = excessive	7 = very poor[a]
Water table depth	1 = 0.1 ft	6 = over 5 ft[a]
Bedrock depth	1 = under 1 ft	7 = over 10 ft[a]
Slope percent	0 = level	6 = 30–45%[a]
Topography		
Distance to water	Euclidian, # of 100-m units	

[a] 9 = variable.

same distribution of prairie, savanna, and forest. Likewise, the land use category of each cell for 1882, 1937, and 1978 was associated with its site type to assess whether deforestation was site selective. If deforestation were not site selective, it would have affected areas in each site type in proportion to its area and the percent of deforested land for the township. Chi-square analysis was used to test for site selectivity of deforestation.

8.2.4 Field Survey of Forest Tracts

We undertook a systematic field survey of all forest tracts in Cadiz Township and western Clarno Township (Fig. 8.1). All but two tracts are privately owned; most are farm woodlots. Vegetation was sampled during summer (May to August) 1983 and 1984. Our goal was to sample one or more cells in each forest tract in Cadiz Township and the western half of Clarno Township. The number of cells sampled in a tract depended on its size, topographic and vegetational heterogeneity, and number of owners. Most forest tracts, except for those in northwest Cadiz Township, have been surveyed. The vegetation data base includes data from 1700 plots in 380 cells located in 98 forest tracts.

Vegetation was surveyed by the stratified random line strip method (Lindsey 1955). Plots were located along 100-m transects across each 1.0-hectare forest cell, beginning at the forest edge. Either one or two transects were sampled in each cell, depending on its homogeneity. Where a transect paralleled a forest edge, it was at least 25 m from the edge. Thus, we sampled both edge and interior, with emphasis on the interior of forest tracts. The groundlayer, shrub, sapling, and tree strata were surveyed using nested plots. Trees (≥ 10 cm dbh) and saplings (2.5 to 9.9 cm dbh) were recorded by species and dbh in 10m × 25-m plots. Species designations follow Gleason and Cronquist (1963). Only tree data are used in the analyses reported here. Data have been summarized for each forest tract.

Table 8.2. Comparison of Tree Importance Values in Presettlement and Present Day Cadiz Township, Wisconsin, East and West of the Pecatonica River[a]

Species Name	Presettlement		Present	
	West	East	West	East
Quercus alba	44.2	34.0	12.0	4.6
Quercus macrocarpa	39.8	10.6	10.1	1.6
Quercus rubra	4.1	16.7	8.8	5.4
Quercus velutina	10.0	4.2	3.7	<0.1
Celtis occidentalis		0.4	6.8	2.4
Acer saccharum		3.4	1.4	28.2
Tilia americana		6.9	4.0	11.4
Ulmus americana		13.2[b]	7.6	7.2
Ulmus rubra			13.6	7.4
Prunus serotina		1.0	10.6	3.6
Fraxinus americana		2.2	1.0	10.4

[a] The west side was previously savanna; forest dominated Cadiz Township east of the river.
[b] Importance value for *U. americana* and *U. rubra*, combined.

8.2.5 Analysis of Field Data

Detrended correspondence analysis (DCA; Hill 1979; Gauch 1982) was used to display compositional relationships among stands. Downweighting was used to reduce the role of rare species in the analysis. Axis scores of stands were also used to construct three-dimensional "maps" of the region. That is, stands were located on a regional map (two dimensions). The third dimension reflected the ordination scores of stands on either ordination axis. Densities of selected species were mapped and displayed by the same method. The SURF routine in the Golden Graphics System (Golden Software 1985) was used to produce these trend surface diagrams. The kriging algorithm was used for trend surface analysis (Ripley 1981).

8.3   Results

8.3.1   Presettlement Vegetation

Wayne, Cadiz, and Clarno Townships had a mosaic of vegetation types characteristic of southern Wisconsin (Fig. 8.2). Prairie and savanna extended from Wayne Township into Cadiz Township, and a ribbon of prairie extended northeast from Illinois into southeastern Cadiz and Clarno Townships. Xeric forest, dominated by *Quercus alba*, *Q. rubra*, *Q. velutina*, and *Q. macrocarpa*, extended across northern Wayne Township into Cadiz Township. Dry mesic forest occupied the core of Cadiz Township and was separated from savanna or prairie by xeric forest. Dry mesic forest abutted prairie directly in eastern Clarno Township. Sedge meadow and marsh were present along the upper Pecatonica River in northwestern Cadiz Township and in northern Wayne Township, and separated the savanna from xeric forest. The lower Pecatonica River separated savanna on the west from xeric forest to the east.

The oaks were the dominant species both east and west of the Pecatonica River (i.e., in both savanna and xeric and dry mesic forest); however, the mix of species differed. *Quercus alba* was the major species in both savanna and forests. *Quercus macrocarpa* was of virtually equal importance west of the river, and it was present east of the Pecatonica River as well but at a lower importance value. *Quercus rubra* was more important to the east and *Q. velutina* to the west. Some species are notably restricted. Neither *Acer saccharum*, *Ulmus spp.*, nor *Fraxinus americana* was recorded as a witness tree west of the Pecatonica River, and only as occasional trees in the forest areas to the east.

8.3.2 Site Versus Presettlement Vegetation in Cadiz Township

Did the vegetation types, especially the distinctions between prairie, savanna, and forest, reflect differences in site conditions east and west of the Pecatonica River? The location of the savanna-forest boundary along the Pecatonica River argues for the importance of the river and its adjacent lowland marsh and sedge vegetation as fire breaks. However, we tested this by deriving a site classification based on soil and topographic data for each cell in the spatial data base of Cadiz Township and relating this to the major vegetation types.

The ten site types are ordered along a topographic gradient (Table 8.3). Five types distinguish sites in valley bottoms according to soil type (including distinctions between organic and mineral soils), drainage, and available soil moisture. Sites on lower and middle slopes are distinguished on the basis of soil depth and slope. One category, "steep and rocky," is both droughty and steep. Middle and upper slopes are distinguished from broad uplands by their lesser soil depth and greater droughtiness.

The distribution of vegetation types among sites (Table 8.4) suggests that there was little topographic or edaphic control over vegetation (marsh, an obvious exception, was excluded from this analysis). A chi-square test suggests that, for site types comprising most of the area in Cadiz Township, the area of the vegetation types does not differ significantly from that expected if each site type had the same proportion of each vegetation type as found in Cadiz Township as a whole. In strict terms, the null hypothesis is not rejected ($P > 0.05$) for the two upland sites and droughty lower-middle slopes. It is rejected for the mesic lower slopes and well-drained bottomland areas. The chi-square test was not applied to sites having fewer than five cells in one or more vegetation types, but only the organic soils in bottomlands lack these vegetation types. However, a caveat is necessary. The site analysis is based on soil and topographic data that were obtained for each 1.0-hectare cell in Cadiz Township. The presettlement vegetation type assigned to each cell is based on a map which is generalized for a region. The map (see Fig. 8.2) does not show possible interdigitation of vegetation types, small enclaves of one type surrounded by others, or a mosaic of vegetation types related to topography. Given this lim-

Table 8.3. Summary (means ± 1 SD) of Variables Used in Cluster Analysis, by Cluster

Topographic Position	Site Type D	Area (hectares)	Texture A Horizon	Depth A Horizon	Available Soil Moisture	Drainage Class	Water Table Depth	Bedrock Depth	Slope Percent	Distance to Water
Bottomland	Muck soils	7	NA NA	NA NA	7.0 0	6.0 0	1.0 0	5.0 0	1.0 0	1.8 0.7
	Peaty muck	10	NA NA	NA NA	20.0 0	1.0 0	1.0 0	5.0 0	1.0 0	3.0 1.3
	Well-drained	500	11.0 0	60.0 0.7	13.0 0.5	3.6 0.5	6.0 0	5.0 0	1.1 0.4	1.6 1.0
	Poorly drained	201	11.0 0	54.0 0	13.0 0	6.0 0	1.0 0	5.0 0	1.0 0	1.8 1.1
Low benches	Low benches	713	10.9 0.5	16.2 1.2	8.3 1.2	5.4 1.1	2.0 1.9	4.7 0.7	1.3 0.6	3.1 1.8
Lower slopes	Lower-middle	1825	10.9 0.6	7.7 1.5	7.3 1.1	4.0 1.3	5.3 1.3	4.4 0.9	2.5 0.9	2.7 1.3
	L-M drouthy	1944	10.5 1.5	5.7 1.6	3.1 1.3	2.9 0.3	6.0 0	2.6 1.2	3.7 1.0	3.0 1.5
Uplands	Steep & rocky	106	99	0.0 0.0	0	9.0 0.0	7.8 0.3	9.0 0	6.6 3.7	2.8 1.7
	Middle-upper	949	10.7 1.0	8.1 1.7	5.6 1.5	3.2 0.7	6.0 0.2	3.5 1.1	2.9 0.8	6.1 1.6
	Broad ridges	3329	11.0 0.1	11.1 1.5	11.2 1.5	3.1 0.4	5.8 0.6	4.6 0.5	2.5 0.7	3.7 1.8

Table 8.4. Distribution of Presettlement Vegetation Types Among Sites in Cadiz Township, Wisconsin, Excluding Area of Marsh[a]

	Uplands		Lower-Middle Slopes			Bottomland					Total Area (ha)
	Broad Ridges	Upper Slopes	Steep Rocky	Dry	Mesic	Low Benches	Poorly Drained	Well-Drained	Mucky Peat	Muck	
Prairie	164	33	0	15	101	1	17	13	0	0	344
Savanna	263	68	3	178	115	9	4	42	0	0	682
Xeric forest	2216	708	84	1563	1051	170	15	167	0	0	5974
Dry forest	501	84	2	148	256	22	16	34	0	0	1063
Total	3144	893	89	1904	1523	202	52	256	0	0	8063
χ^2 test (0.05)[b]	A			A				R			

[a] Hectare cells.

[b] A = accept, R = reject null hypothesis.

itation in resolution, we conclude that site factors are not the primary determinants of the location of the major vegetation regions. The Pecatonica River and its adjacent wetlands functioned as a fire barrier, protecting the areas in Cadiz Township east of the river from the fires that maintained the prairie and savanna to the west.

8.3.3 Deforestation Regimes

As settlement began in the mid-1800s, about 80 to 85% of Cadiz Township was covered by forest and savanna. Deforestation was rapid (Tables 8.5 and 8.6). Forest had declined to about 30% by 1882, and now is about 10% of the area. Concurrently, the number of separate forest tracts increased, to about 63 by 1882. As deforestation continued, some of these were removed, whereas others were reduced in size or dissected to become two or more smaller parcels. The number and total area of parcels has been stable for the past half-century; some deforestation has been counteracted by reversion of limited areas to forest.

This deforestation regime has created a large number of isolated forest tracts (see Fig. 8.1). Whereas the presettlement vegetation units were largely contiguous, most farm woodlots are surrounded by agricultural land and separated from nearest neighbors by hundreds of meters. Fencerows connect some of these forest tracts, but most fencerows are well maintained and support little natural vegetation.

The impact of the rectangular land survey system is apparent in the geometry of farms, fields, and woodlots. No farms lack a woodlot, and few have more than one. Most are as rectilinear as the fields and are subunits of the land survey system. In fact, our 1.0-hectare grid system fits precisely over most forest tracts; for example, a 10-acre woodlot is a 4.0-hectare block. It is clear that the settlers followed the dictates of the land survey system to create ``order upon the land'' (Johnson 1976). Did they disregard site differences in allocating some land to forest such that these forest tracts represent a cross section of site types? Or, conversely, have some site types been deforested selectively? The deforestation regime is shown for the major site types in Table 8.5, and graphically in Fig. 8.3 for six major site types. The areas of each type in Fig. 8.3 are standardized to highlight deviations in deforestation rates for each site from the average rate for Cadiz Township. We tested the null hypothesis that residual forest in each site type was the same as for Cadiz Township as a whole using the chi-square test. Results of this test are translated into whether forest area was above average, average, or below average in area in 1882, 1937, and 1978.

It is clear that distinctions were made during early stages of deforestation, and that distinctions continue to be made. By 1882, the broad ridges, upper slopes, and lower-middle slope sites, except for steep and rocky areas, were significantly deforested. Low benches and bottomlands were partially deforested, but had above-average forest areas. Settlers primarily cleared uplands during early stages of agricultural development. This pattern of deforestation of the uplands continued from 1882 to 1937, when some of the sites that had

Table 8.5. Deforestation Regime by Site Type in Cadiz Township, Wisconsin (Savanna, Xeric Forest, and Dry Mesic forest are included)[a]

Period	Uplands		Lower-Middle Slopes				Bottomland			
	Broad Ridges	Upper Slopes	Steep Rocky	Dry	Mesic	Low Benches	Poorly Drained	Well-Drained	Mucky Peat	Muck
Presettlement	2980	860	89	1885	1422	201	35	243	0	0
1833–1882 Rate	Avg.	Above	Below	Above	Avg.	Below	Below	Below		
1882 Area	1281	312	54	638	630	198	54	141		
x^2 Test (0.05)	Avg.	Below	Above	Below	Avg.	Above	Above	Above		
1882–1937 Rate	Above	Avg.	(Refor)	Zero	Above	Avg.	(Refor)	Zero		
1937 Area	505	174	65	637	343	90	63	145		
x^2 Test (0.05)	Below	Below	Above	Above	Below	Above	Above	Above		
1937–1978	Avg.	Zero	Zero	(Refor)	Above	Above	Avg.	Avg.		
1978 Area	487	176	67	750	286	60	56	138		
x^2 Test (0.05)	Below	Below	Above	Above	Below	Avg.	Above	Above		

[a] Chi-square test ($P < 0.05$) used to test hypothesis that deforestation of a site type was same as regional average.

Table 8.6. Trends in Number, Average Size, and Isolation of Forest Tracts in Cadiz Township, Wisconsin

Year	Number of Tracts	Total Area (ha)	Average size (ha)	Isolation	
				Average (m)	Extreme (m)
1833		8660			
1882	63[a]	2703	43	202	800
1902	44[a]	909	21		
1937	97	784	12	250	1200
1963	117	925	8	211	1200
1978	111	1000	9	207	1200

[a] Based on Shriner and Copeland (1904). Includes only forest tracts ≥ 4 hectares.

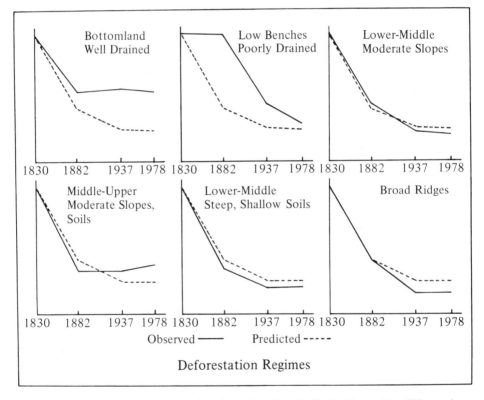

Figure 8.3. Deforestation regime for six major sites in Cadiz Township, Wisconsin, from before agricultural settlement (1833) to 1978.

been neglected in the early stage of deforestation had above-average deforestation rates. Notably, deforestation of bottomlands and areas of rough topography virtually stopped. Currently, the mesic and well-drained uplands have below average forest areas, whereas the droughty lower-middle slopes and both well-drained and poorly drained bottomlands have above-average forest area. Only the low benches have a ratio of forest to nonforest that equals that of Cadiz Township as a whole.

Although no site has been deforested completely, a great deal of selectivity has been practiced. This agrees with Auclair's (1976) findings that site distinctions are being made in land use decisions in southern Wisconsin, but the segregation among land uses is not as clear-cut as in his study area. Our field observations suggest that much of the forested area that might be considered suitable for pasture or tillage is in a block of forest that includes areas of rougher topography. In other words, forest areas generally have a size and shape that is derived from the land survey system, but a location that expresses topographic and edaphic considerations. Consequently, the remnant forest patches tend to be topographically diverse. Because remnant forest has been left on a representative sample of sites, even though in uneven proportions, this suggests that the deforestation regime, per se, has not tended to cause the loss of tree species from this landscape.

8.3.4 Tree Species Composition

Each woodlot has a species composition and stand structure that are derived from the presettlement forest, postsettlement patterns of disturbance (primarily anthropogenic), and ongoing species replacement patterns. Our survey of each woodlot in Cadiz and western Clarno Townships gives a spatial representation of the forests of this landscape.

A number of analyses can be made to describe the spatial distribution of the species. At the regional scale, current species importance values can be compared with those of the presettlement forest reconstructed from witness tree data (Table 8.2). Two patterns stand out. First, there has been a marked and general decline in the oaks. For example, the importance value of *Q. alba* trees has declined from 34.0 east of the Pecatonica River in 1833 to 4.6 in the 1980s. Only *Q. rubra* seemingly has increased in importance, on the west side of the river. Concomitantly, the importance values of several mesic species have increased. Most notable is *Acer saccharum*, whose importance east of the river has increased from 3.4 to 28.2. The *A. saccharum* witness trees were widely scattered in the dry mesic forest area. Now virtually all woodlots east of the Pecatonica have significant amounts of *A. saccharum* canopy trees. The incidence of *A. saccharum* west of the river is much more limited, however, suggesting that it has not invaded these woodlots extensively. The importance of *Tilia americana* and *Fraxinus americana* also has increased, again primarily east of the river. There are two possible interpretations of these increases in species importance. Either these species have dispersed from the occasional individual trees or stands in the presettlement forest that were outside the boundaries of the current remnant forest areas, or they have increased in importance without significant dispersal as a result of management practices.

Detrended correspondence analysis (DCA) ordination gives a more detailed analysis of the vegetation patterns in Cadiz and Clarno Townships. The first two DCA axes gave clearly interpretable patterns of vegetation. The first ordination axis shows a trend from forest tracts dominated by *A. saccharum* to oak-dominated stands (Fig. 8.4A). The second axis shows a separation of *Q. macrocarpa* dominated stands from stands dominated by *Q. alba* and *Q. rubra*.

Mapping of the ordination axis scores and of the density of selected species (Fig. 8.4B-F) shows spatial patterns that help interpret the vegetation of these townships. Each three-dimensional diagram is a view of Cadiz and western Clarno Townships, as viewed from the southwest corner of Cadiz township, with an observation angle of 45° above the horizon. The trend surface is derived by considering each forest tract as a point sample, and applying trend surface analysis to interpret between the points.

Mapping of the first ordination axis (Fig. 8.4B) shows that stands in the southern and central portions of the area generally have low axis scores. Scores increase toward the west, east, and north. That is, stands dominated by *A. saccharum* are localized in the south-central portion of the area, whereas oak-dominated stands are to the west, north, and to a lesser extent, to the east.

The three-dimensional map of the second ordination axis (Fig. 8.4C) places the trend from *Q. alba* to *Q. macrocarpa* in geographic context. Intermediate values are concentrated in the central and eastern part of the area. These values represent stands dominated by *A. saccharum* and other mesic species (e.g., *T. americana*). There is a sharp decrease in axis score to the west, and an increase to the north and northeast; that is, there is a spatial distinction between *Q. macrocarpa* and *Q. alba*.

Mapping the density of selected species clarifies the distribution of axis scores. *Acer saccharum* densities are highest in the central part of the area (Fig. 8.4D). They decline eastward. They also decline to low values westward and are negligible west of the Pecatonica River. *Quercus alba* densities are low in southern Cadiz and Clarno Townships, but they increase rapidly toward their northern border (Fig. 8.4E). The distribution of *Q. macrocarpa* (Fig. 8.4F) is especially striking. It is highest in northwest Cadiz township and generally high

---

Figure 8.4. Ordination and three-dimensional maps of first and second ordination axes and of tree densities based on all trees (dbh ⩾ 10 cm) measured in 92 forest tracts in Cadiz and Clarno Townships, Wisconsin. (A) Detrended correspondence analysis; (B) three-dimensional map of first ordination axis scores; (C) three-dimensional map of second ordination axis scores; (D) three-dimensional map of densities of *Acer saccharum* trees; (E) three-dimensional map of densities of *Quercus alba* trees; (F) three-dimensional map of densities of *Quercus macrocarpa* trees. All three-dimensional maps are of Cadiz and western Clarno Townships as viewed from over the southwest corner of Cadiz Township, and at an angle of 45° above the horizon.

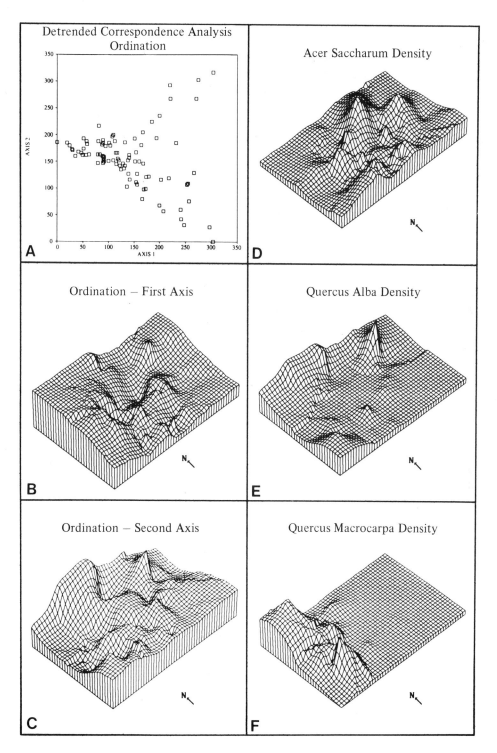

along its western boundary. Only three stands east of the Pecatonica River have *Q. macrocarpa*, and these are in low densities as shown by the small "hills" in the generally flat surface in the eastern three-quarters of the area.

Mapping of ordination scores and densities of *A. saccharum*, *Q. macrocarpa*, and *Q. alba* suggests that the general spatial pattern of the vegetation has not changed markedly since settlement began (compare Fig. 8.2 and 8.4). *Acer saccharum* is located in the areas identified as having dry mesic forest in the presettlement landscape. *Quercus alba* generally has its highest densities in areas formerly occupied by xeric forest. *Quercus macrocarpa* is virtually limited to woodlots that have been derived from presettlement savanna.

8.4   Discussion and Conclusions

Past ecologic research leads us to expect that: (1) the pattern of presettlement vegetation was closely tied to the pattern of disturbance, especially by fire (Anderson and Anderson 1975; Grimm 1984); (2) site differences in uplands affected vegetation pattern primarily through their influence on disturbance regime (Grimm 1984); (3) removal of the presettlement disturbance regime, especially large-scale fire, leads to convergence in structure and species composition in forests and an increase in the importance of such mesic species as *A. saccharum* (Bray and Curtis 1957; Curtis 1959; Peet and Loucks 1977); and (4) the degree of this convergence, and the species diversity in the collection of stands, is mediated by succession and site conditions of the respective stands (Peet and Loucks 1977). Our findings agree with the first two generalities, but the organization of the landscape as it affects species dispersal appears to operate as a factor in the latter two in ways that have not been recognized.

It is clear that *A. saccharum* and other mesic species have increased in importance. However, these species are still not well represented in new locations west of the Pecatonica River (Table 8.2, Fig. 8.4D). By contrast, most tracts east of the river in areas formerly occupied by xeric and dry mesic forest are now dominated by mesic species. Apparently, these forests are undergoing the postfire succession that has been considered characteristic of southern Wisconsin (Bray and Curtis 1957; Peet and Loucks 1977). Whether mesic species have invaded these woodlots or have benefited from competitive release is not clear.

We might attribute the current, but localized, presence of *A. saccharum* in a few woodlots west of the Pecatonica River to dispersal across this former fire barrier. Conversely, the absence of *A. saccharum*, as well as *T. americana*, and *Fraxinus spp.*, from most woodlots west of the Pecatonica River suggests that such landscape-scale dispersal has not taken place. These wind-dispersed species generally do not disperse effectively more than several hundred meters, even over open terrain such as fields (Sharpe and Fields 1982). Furthermore, those forest tracts in western Cadiz Township in which *A. saccharum* is found include mesic sites in small valleys or bluffs along the west bank of the Pecatonica River, which may have served as refugia in this generally fire-disturbed presettlement landscape. Additionally, mesic sites were not cleared as rapidly as upland sites (e.g., low benches were not cleared in early stages of settlement; see Fig. 8.3), and these may have served as local seed sources for some forest tracts west of the river before they were cleared. In general, the speed of deforestation and isolation of seed sources may have been too rapid for species with these limited dispersal capabilities to disperse widely before the remnant forest was too fragmented and isolated for effective dispersal between forest tracts. Remnant forest distant from seed sources for mesic species in the presettlement landscape seem not to have acquired them through long-distance dispersal.

Acer saccharum seems to be increasing in importance in many other Midwestern oak-dominated forests (Rochow 1972; Schlesinger 1976; Miceli et al. 1977). Nigh et al. (1985a,b) remeasured permanent plots in upland oak-hickory forests in central Missouri, in an area where presettlement records indicated that *A. saccharum* was largely confined to relatively isolated sites (Howell and Kucera 1956). Although *Quercus* is still the dominant canopy tree in these forests, *A. saccharum* increased in importance in both sapling and canopy layers in a 14-year period (1968 to 1982). The Missouri forests studied by Nigh et al. (1985a,b) were in state forests and other public forest areas, and may not be as isolated, historically or currently, as the stands in the south-central Wisconsin agricultural landscape that we have studied. Brownfield Woods, studied by Miceli et al. (1977), has an area of 24 hectares. The stand in the Kaskaskia Experimental Forest in southern Illinois studied by Schlesinger (1976) is contiguous to an extensive forest area. Invasion by *A. saccharum* may depend on the proximity of seed sources that survive postsettlement deforestation. There may be threshold levels of isolation beyond which invasion will not take place. Consequently, we would expect that *A. saccharum* dispersal is slower in agricultural landscapes with isolated woodlots, and that the presettlement patterns of species distribution are likely to be retained longer.

The current woodlots have a tree species composition that is a direct legacy of their locations with respect to species distribution in the presettlement landscape. For each tract, succession may be limited by its presettlement species composition and that of adjacent tracts now deforested, and by postsettlement species removal by disturbance, analogous to the limits placed on succession by site limitations (Peet and Loucks 1977). These limits are not solely intrinsic to the site, but are imposed by the spatial organization of the presettlement forest and of the present forest tracts. Consequently, each woodlot has a species composition and stand structure derived from the presettlement forest (and its natural disturbance regime), the postsettlement patterns of disturbance (primarily anthropogenic), and ongoing species replacement patterns. Explanation of species composition and replacement patterns in individual stands must take into account stand location, both in the current landscape and historically. This idea merits further study, particularly in light of recent attempts to use isolated woodlots as "islands" in island biogeographic studies (Tramer and Suhrweir 1975; Levenson 1981; Weaver and Kellman 1981).

Acknowledgments

The efforts of many people have made possible this and other parts of our landscape ecology research program. We are particularly indebted to a willing corps of field assistants who helped us during the 1983 and 1984 field seasons, and laboratory assistants who helped us develop the spatial data base of Cadiz Township. Our field station was provided by the Green County Forestry Education Association, and we thank Mr. Albert Deppeler and Ray Amiel, Wisconsin Department of Conservation, for their assistance and advice.

References

Anderson, R.C., Anderson, M.R. 1975. The presettlement vegetation of Williamson County, Illinois. *Castanea* 40:345–363.

Auclair, A.N. 1976. Ecological factors in the development of intensive management ecosystems in the Midwestern United States. *Ecology* 57:431–444.

Auclair, A.N., Cottam, G. 1971. Dynamics of black cherry (*Prunus serotina* Erhr.) in southern Wisconsin oak forests. *Ecol. Monogr.* 41:153–177.

Bray, J.R., Curtis, J.T. 1957. An ordination of the upland forest communities of southern Wisconsin. *Ecol. Monogr.* 27:325–348.

Curtis, J.T. 1956. The modification of mid-latitude grasslands and forests by man, pp. 721–726. *In* W.L. Thomas (ed.), *Man's Role in Changing the Face of the Earth.* University of Chicago Press, Chicago, IL.

Curtis, J.T. 1959. The vegetation of Wisconsin: An ordination of plant communities. University of Wisconsin Press, Madison, WI.

Curtis, J.T., McIntosh, R.P. 1951. An upland forest continuum in the prairie-forest border region of Wisconsin. *Ecology* 32:476–496.

Davis, A.M. 1977. The prairie-deciduous forest ecotone in the Upper Middle West. *Ann. Assoc. Am. Geogr.* 67:204–213.

Finley, R.W. 1976. Original Vegetation Cover of Wisconsin from U.S. General Land Office Notes (map). U.S. Forest Service, North Central Forest Experiment Station, St Paul, MN.

Gauch, H.G. Jr. 1982. *Multivariate Analysis in Community Ecology.* Cambridge University Press, New York.

Gleason, H.A., Cronquist, A. 1963. *Manual of Vascular Plants of Northeastern United States and Adjacent Canada.* Van Nostrand, New York.

Glocker, C.L. 1974. *Soil Survey of Green County, Wisconsin.* U.S. Dept. Agric. Soil Conserv. Serv., Washington, D.C.

Golden Software. 1985. Golden graphics system. Golden, CO.

Grimm, E.C. 1984. Fire and other factors controlling the Big Woods vegetation of Minnesota in the mid-nineteenth century. *Ecol. Monogr.* 54:291–311.

Harper, K.T. 1963. Structure and dynamics of the maple-basswood forests of southern Wisconsin. Doctoral thesis, University of Wisconsin, Madison, WI.

Hill, M.O. 1979. *DECORANA: a FORTRAN Program for Detrended Correspondence Analysis.* Cornell University, Ithaca, NY.

Howell, D.L., Kucera, C.L. 1956. Composition of presettlement forests in three counties of Missouri. *Bull. Torrey Bot. Club* 83:207–217.

Johnson, H.B. 1976. *Order Upon the Land: The U.S. Rectangular Land Survey and the Upper Mississippi Country.* Oxford University Press, New York.

Kline, V.M., Cottam, G. 1979. Vegetation response to climate and fire in the Driftless Area of Wisconsin. *Ecology* 60:861–868.

Levenson, J.B. 1981. Woodlots as biogeographic islands in southeastern Wisconsin, pp.

13–39. *In* R.L. Burgess and D.M. Sharpe (eds.), *Forest Island Dynamics in Man-Dominated Landscapes*. Springer-Verlag, New York.

Lindsey, A.A. 1955. Testing the line-strip method against full tallies in diverse forest types. *Ecology* 36:485–495.

Miceli, J.C., Rolfe, G.L., Pelz, C.D.R., Edgington, J.M. 1977. Brownfield Woods, Illinois: Woody vegetation changes since 1960. *Am. Midl. Nat.* 98:469–476.

Nigh, T.A., Pallardy, S.G., Garrett, H.E. 1985a. Sugar maple-environment relationships in the River Hills and Central Ozark Mountains of Missouri. *Am. Midl. Nat.* 114(2):235–251.

Nigh, T.A., Pallardy, S.G., Garrett, H.E. 1985b. Changes in upland oak-hickory forests in central Missouri, 1968–1982, pp. 170–177. *In* J.O. Dawson and K. Majerns (eds.), *Proceedings of the Fifth Central Hardwood Forest Conference*. April, 1985, Champaign-Urbana, IL.

Peet, R.K., Loucks, O.L. 1977. A gradient analysis of southern Wisconsin forests. *Ecology* 58:485–499.

Radloff, D.L., Betters, D.R. 1978. Multivariate analysis of physical site data for wildland classification. *For. Sci.* 24:2–10.

Ripley, B.D. 1981. *Spatial Statistics*. Wiley-Interscience, New York.

Rochow, J.J. 1972. A vegetational description of a mid-Missouri forest using gradient analysis techniques. *Am. Midl. Nat.* 87:377–396.

Rogers, D.J. 1959. Ecological effects of cutting in southern Wisconsin. Doctoral thesis, University of Wisconsin, Madison, WI.

SAS. 1985. *SAS User's Guide: Statistics Version 5*. SAS Institute Inc, Cary, IN.

Schlesinger, R.C. 1976. Hard maples increasing in upland hardwood stand, pp. 177–185. *In* J.S. Fralish, G.T. Weaver, and R.C. Schlesinger (eds.), *Proceedings of the First Central Hardwood Forest Conference*. Southern Illinois University, Carbondale, IL.

Sharpe, D.M., Fields, D.E. 1982. Integrating the effects of climate and seed fall velocities on seed dispersal by wind: A model and application. *Ecol. Model* 17:297–310.

Shriner, F.A., Copeland, E.B. 1904. Deforestation and creek flow about Monroe, Wisconsin. *Bot. Gaz.* 37:139–143.

Sinton, D.F. 1977. *The User's Guide to I.M.G.R.I.D.: An Information Manipulation System for Grid Cell Data Structures*. Harvard University, Dept. Landscape Architecture, Graduate School of Design, Cambridge, MA.

Tomlin, C.D. 1983. Digital cartographic modeling techniques in environmental planning. Doctoral dissertation, Yale University, School of Forestry and Environmental Studies, New Haven, CT.

Tramer, E.J., Suhrweir, D.E. 1975. Farm woodlots as biogeographic islands: Regulation of tree species richness. *Bull. Ecol. Soc. Am.* 56:53.

Weaver, M., Kellman, M. 1981. The effects of forest fragmentation on woodlot tree biota in southern Ontario. *J. Biogeogr.* 8:199–210.

3. Implications for Landscape Management

9. Landscape Restoration in Response to Previous Disturbance

Darrel Morrison

9.1   Introduction

The preceding chapters have treated the landscape as an object that is acted upon by a disturbance. It is of equal interest to consider how landscape heterogeneity can be managed so that the consequences of some disturbances can be corrected.

One approach to landscape management that has the potential of increasing or restoring natural species diversity and aesthetic character to previously disturbed sites is the re-establishment of assemblages of native plant species in groupings that resemble the plant communities that might have occupied those sites before disturbance. There is a wide spectrum of scales at which landscape restoration might be practiced—ranging from small "native gardens" planted in residential backyards and corners of parks and botanical gardens to large restorations of prairie, forest, marsh, or dune vegetation that may occupy tens of hectares.

The purpose of this chapter is to provide (1) a practical definition and a philosophy of landscape restoration; (2) a review of selected historical examples of restoration; (3) a discussion of the use of naturally evolved landscapes as models for restoration; (4) theoretical and practical considerations in developing a restoration strategy; and (5) aesthetic characteristics of naturally evolved landscapes that may be emulated in restorations.

The following discussion of the practice of landscape restoration emanates

my attempts as a landscape architect to re-establish community-
f native species to Midwestern United States sites. These efforts
th in size and in "authenticity," that is, fidelity to the compo-
ristics of the naturally evolved models on which they are based.

9.2   Definitions and a Philosophy of Landscape Restoration

The term "landscape restoration" implies returning a site to some previous
state, with the species richness and diversity and physical, biological, and aesthetic characteristics of that site before human settlement and the accompanying
disturbances. This ideal is a practical impossibility in most if not all landscape
situations due to the effects of previous and present disturbances.

A more realistic definition of landscape restoration is the following: The
reintroduction and re-establishment of community-like groupings of native species to sites which can reasonably be expected to sustain them, with the resultant
vegetation demonstrating aesthetic and dynamic characteristics of the natural
communities on which they are based.

The goals and objectives for restoring landscapes may vary. In one case,
the goal may be primarily functional, such as providing a diverse natural cover
of vegetation for erosion control or for wildlife habitat. In another, the goal
may be primarily aesthetic, such as providing the appearance of a tall-grass
prairie or freshwater marsh. Realistically, most restorations will be, at least in
the beginning, somewhat simplified versions of the naturally evolved models.
But such scaled-down restorations can, over time, closely resemble the composition and structure of the landscapes after which they are modeled. This
requires the presence of one or more of the following conditions:

1. A reservoir of propagules of endemic species remaining on or near the restoration site, with the potential of successfully invading and colonizing on
   that site, thereby enriching the species composition over time.
2. A relative absence of aggressive or opportunistic exotic species' propagules
   on or near the site, as they might out-compete the native species being re-
   established.
3. A management program that emulates natural phenomena that would have
   normally occurred in the naturally evolved model, for example, fire in prairie
   restorations, or fluctuating water levels in marshes.

The underlying premise of landscape restoration is that there are many situations in the human-dominated landscape where community-like groupings of
native plant species can provide a vegetative cover that is, simultaneously,
functional, economical, ecologically suitable, and aesthetically appealing.
Among the more obvious examples of sites where this might be appropriate
are large parks, roadsides and utility corridors, "greenways" in urban areas,
and large commercial or industrial sites. Additionally, certain zones might be
set aside on predominantly agricultural sites, where a naturally diverse complex
of native vegetation might serve such purposes as erosion and sedimentation

control and/or soil improvement in a long-term "crop rotation." Not to be overlooked for potential restoration is some percentage of the ubiquitous lawns surrounding suburban residences, which cumulatively require an estimated 40 million lawnmowers using 200 million gallons of gasoline annually, in addition to massive inputs of water, fertilizer, and herbicides (Diekelmann and Schuster 1982).

At its best, landscape restoration may lead to the re-establishment of communities of plants and animals that have been greatly diminished by human activities. It may even provide hospitable environments for populations of plants or animals that are rare and endangered. But our ability to restore natural communities to some extent should never be seen as a substitute for the *preservation* of extant surviving examples of natural communities. Whenever landscape restoration practices have the effect of damaging naturally evolved landscapes (e.g., through removing plants to be transplanted to the restored versions), they should be questioned. Similarly, if the possibility of partial restoration of a community leads to an attitude that diminishes the value placed on the "original," restoration should be questioned.

In fact, the practice of landscape restoration should lead to our placing greater value on naturally derived plant communities, because the composition and structure of such communities within close geographic proximity of a site provide the best basis for planning a community restoration on that site. This is true, even if one is attempting a restoration that has the primary goal of emulating the *appearance* of the natural community.

9.3   Historical Examples of Landscape Restoration Activity

Although such activities as prairie and marsh restoration are often thought of as relatively recent ventures in landscape design and management, there are historic examples in which restoration of groupings of native plants was the central theme (Morrison 1981).

For example, in 1888, Jens Jensen, Danish-born landscape architect for the Chicago park system, designed and installed the "American Garden" in the corner of Washington Park, using wildflowers which were fast disappearing from the urbanizing environment of Chicago (Eaton 1964). This small but symbolic effort was followed by a half-century career in which Jensen increasingly emulated natural plant communities' composition and dynamics in his designed landscapes, both public and private. It is noteworthy that he purposefully observed natural areas in Illinois and Wisconsin to learn the plant species and understand their distribution patterns and interrelations with each other and their environment. It is also noteworthy that a frequent associate on field trips was the botanist-ecologist H. C. Cowles. One of the best examples of Jensen's artistic landscape restoration ability was his "prairie river" landscape in Columbus Park. In it, he successfully combined mowed "meadows" as activity areas, with a curving swath of prairie and river-edge vegetation alongside a lagoon, backed up by rounding masses of native shrubs and trees. Although this specific combination of forms and plant species might never have occurred on this site, it constituted an aesthetically pleasing, ecologically suitable complex that looked like it might have evolved naturally. Jensen's philosophy on the beauty of natural landscapes and the value of preserving them for study—as well as for their own intrinsic value—is compiled in his book, *Siftings* (Jensen 1939).

In 1929, another landscape architect, Elsa Rehmann, teamed up with a botanist, Dr. Edith A. Roberts, to produce a book entitled *American Plants for American Gardens* (Roberts and Rehmann 1929). In it, they advocated observation of native plant communities as a basis for planting residential sites. In the book, each chapter features species lists and a summary of the environmental and aesthetic characteristics of such northeastern U.S. communities as "The Juniper Hillside," "The Oak Woods," and "The Hemlock Ravine," with suggestions on siting a house and integrating community groupings of native species in site planning and design. There was no attempt to provide propagation or establishment information.

In 1932, three years after the publication of the Roberts and Rehmann book, B.W. Wells, a botany professor at North Carolina State College, published *The Natural Gardens of North Carolina* (Wells 1932). He, like Roberts and Rehmann, included chapters on a variety of diverse plant communities, but focused on the State of North Carolina, discussing the environment, species composition, and descriptions, as well as the communities' visual characteristics, and sometimes the effects of perturbations like fire and flooding. Lists of native species for planting in specific habitats were also provided.

Although there is not evidence that these and similar authors and designers had a widespread impact on landscape architecture during the 1930s to 1950s, landscape restoration was taken up by others, notably botanists (Morrison 1981). Among the acknowledged leaders in this relatively new activity were John T. Curtis and Henry C. Greene of the University of Wisconsin Arboretum, whose prairie restoration attempts began in 1935. By the early 1950s, they reported that it was abundantly clear that such restoration could be successful (Curtis and Greene 1953). Logically, their restoration work began with thorough studies of remnant natural prairies. The two restored prairies in the University of Wisconsin Arboretum today are testimony to their success. Greene Prairie, in particular, so closely resembles true prairie in its matching of species with habitat and in overall patterns that most observers think it is a purely natural phenomenon. Greene's careful records of planting locations and techniques in this project provide baseline data for monitoring the dynamics of the plantings and evaluating the success of installation and management techniques today (Berger 1985).

The University of Wisconsin prairie restoration work encouraged others to follow, such as Ray Schulenberg of the Morton Arboretum during the 1960s (Berger 1985). Subsequently the experience in prairie restoration has, particularly in the last two decades, spread to a variety of other ecosystems, including forests, coastal dunes, and both salt marshes and freshwater marshes (Thorhaug 1980). In addition, there is a growing body of experience and literature in land reclamation, mainly on lands that has been drastically disturbed by coal and metallic mineral extraction processes (Hutnik and Davis 1973; Schaller and Sutton 1979; Bradshaw and Chadwick 1980; Chadwick 1980; Chadwick 1982; Berger 1985). Since 1981, a journal, *Restoration and Management Notes*, edited at the University of Wisconsin Arboretum by William Jordan, III, has documented a large number of restoration-related experiments and projects.

Such historical examples of landscape restoration at many scales and under a wide range of circumstances suggest that restoration is a viable and practicable approach to ameliorate negative effects of earlier human activity. The remainder of this chapter will be devoted to practical considerations in planning for landscape restoration; that is, information that can be obtained from naturally evolved landscapes, factors that influence a site-specific strategy for restoration, and aesthetic characteristics in natural landscapes that can be emulated in restorations.

9.4 "Natural Models" as a Basis for Restoration

Although most restoration projects are, at least initially, greatly simplified versions of the natural landscapes they emulate, in terms of species richness and matching of species with microenvironments, the best primary sources of information are the natural models. Although there are many factors which aren't easily understood through visual analysis of a natural plant community, such as presence or absence of micro-organisms and allelopathic substances in the soil, there are other items of useful information that are clearly observable. For example, there are comprehensive quantitative data on environmental and compositional characteristics for a variety of native plant communities for some geographic areas. One of the best examples of this is John T. Curtis' *Vegetation of Wisconsin* (1959), which includes summary data on more than 20 communities of that state. In all cases, the data for each community type were based on studies of a number of different stands. Hence, the data represent a composite view of the community type, and do not typically portray an individual stand. Nevertheless, for the geographic area covered in these compilations, such information provides an extremely useful starting point for a plant community restorationist.

For areas that do not have such plant community studies, firsthand observation and analysis of extant natural stands can provide the restorationist with not only useful quantitative data, but a variety of other valuable observations as well. In a field course on native plant communities co-taught by Dr. Evelyn Howell, a plant ecologist, and the author at the University of Wisconsin 1973 to 1983, the following mechanisms or exercises were developed to help Landscape Architecture students understand the composition and dynamics of naturally evolved landscapes (Morrison and Howell 1983).

1. Species checklist. On forms prepared for each community type, students note presence or absence of species listed, as well as size classes represented (for trees) and relative abundance.

2. Plant drawings. On those same forms, quick diagnostic sketches of plants' leaves, flowers, or other distinctive characteristics help students remember the identity of plants. On occasion, more carefully drawn "plant portraits" of visually important species of a community give special emphasis to those species.

3. Plot mapping. In areas ranging in size from 1 or 2 square meters (for prairie, sand dune, or forest groundlayer species) up to plots of 100 square meters (for forest trees and shrubs), mapping of plant locations on grid paper provides specific information on density and spatial distribution, some of which may be translated into restoration design.

4. Community sketching. Because a plant community is more than the sum of its parts, another useful exercise is to have students portray the overall characteristics of a plant community, often in fairly quick abstractions of characteristic line, form, color, texture, and pattern. Although the drawings that result from this exercise are excellent memory joggers, the most value probably is derived from students' having to select what they consider to be a representative or characteristic scene to draw, and from the closer observation required in drawing a community's visual characteristics.

5. Aesthetic assessment. In this exercise, students are asked to complete forms that include spaces for such information as descriptive terms for a community's characteristic line, form, color, texture, and spatial qualities, as well as nonvisual attributes. Additionally, space is provided to list plant species that are particularly important to the community's aesthetic character, that is, the "visual essence" species. Such assessments are limited of course, because they identify the important characteristics for the particular day that observations are made, and may overlook ephemeral phenomena like flowering or fall foliage color changes. Nevertheless, like the community sketching exercise, they require closer, more purposeful observations than might otherwise be made.

6. Quantitative sampling. Conventional field sampling exercises also enhance observation and generate quantitative data that can be useful to a restorationist. "Nested" quadrats of different sizes for the different layers of vegetation in a forest, meter- or half-meter-square quadrats in prairie and bog communities, and transect sampling along an environmental gradient are all useful in learning more about naturally evolved plant communities.

John Cairns, in an essay in *Restoration and Management Notes* (1981), logically suggested that ecologists should study disturbed sites—even severely disturbed ones—as a basis for restoration and management. There is no doubt that such observations as those studying the impacts of human activity on the land, the successional trends in disturbed settings, provide useful information for restorationists. Yet, to plan restorations that have compositional characteristics emulating natural communities, there is no substitute for careful observation of preserved naturally evolved ones. Such sites are, in effect, baselines against which the composition and dynamics of disturbed landscapes can be measured. In reality, of course, almost no preserved site can be considered pristine; hence, even field study of preserved natural areas provides opportunities for observing human-induced disturbances.

9.5   Development of a Restoration Strategy

In effect, whenever a restoration project is implemented, a successional process is initiated. There are of course different views, historically, of succession, beginning with Clements' concept in which the initial colonizers of a site are seen to modify it to facilitate the establishment of other species less able to tolerate the harsh or extreme environmental conditions (Clements 1916). Later theorists include Connell and Slatyer (1977), who identified three models of succession: the facilitation model, the tolerance model, and the inhibition model. They stated—and there are examples in the field to support their views—that although Clements' facilitation model does occur, so do other scenarios. Their "tolerance" model suggests that the first colonizers neither facilitate nor impede the establishment of later colonizers, but that their propagules just arrived first. Connell and Slatyer's "inhibition" model, on the other hand, suggests that the first colonizers on a site actually inhibit the establishment of later arrivals on a site. This may result from the fact that the first species to establish may occupy all the available space, as in a sod-forming perennial or a species that is allelopathic.

In planning the initial stages of a restoration, then, the species selection can have a major effect on the dynamics and the long-term species composition of a site.

Other factors that enter into the planning of a restoration include, but are not limited to, the following.

9.5.1   Existing Conditions and "Disturbance History" of the Restoration Site

The presence or absence of an organic soil is a key factor. In the case of land that has been severely modified by mineral extraction processes, there often is not a layer of organic soil, and the restoration plan demands initiation of a primary successional sequence, in which species are selected and soil amendments are added to accelerate the development of a more "natural" soil. On the other hand, on certain oldfield sites, conditions are such that the initial species introduced can be those that are considered mid- or late-successional.

The site will also dictate whether a "complete" or partial restoration is appropriate. If, for example, there is an existing tree canopy, but no trees of smaller size classes and an absence of characteristic shrubs and herbs, the logical approach would be to let the canopy remain and reintroduce the midstory and groundlayer species. The strategy may be less obvious in other situations, where, for example, there might be a mix which includes an aggressive exotic species and small numbers of a relatively rare native species whose presence in the restoration is important. A choice may be necessary, between a difficult selective eradication of just the exotic species or an easier removal of both species.

In general, a restoration site will include plants and/or propagules of exotic species. The abundance of these, and their expected aggressiveness or persistence, will determine the importance of eliminating them as an early step in restoration.

On the other hand, a ready supply of propagules of desirable native species on and adjacent to the site may suggest an approach of introducing an initial cover which could be "tolerant" of invasion and establishment of desirable native species. This initial cover might be an annual, a biennial, or a short-lived perennial, or possibly a nonpersistent species that would be expected to be "phased out" as native species establish and change the environment, for example, by shading the ground.

9.5.2   Restoration Goals and Objectives

As suggested at the outset, the likelihood of implementing a "complete" restoration of a previously disturbed or destroyed plant community is close to nil. Restorations often have a primary goal, and that goal influences species selection and restoration techniques. For example, in a restoration designed primarily to create habitat for certain animal species, plant selection may be skewed toward those that provide appropriate food and cover. For restorations whose principal goal is natural interpretation, the selection and placement of native species may be executed to insure that a wide array of species may be observed from a trail or observation deck. For restoration projects with a central aesthetic goal, planning might focus particularly on the "visual essence" species for the community, in addition to the dominant and prevalent species in the naturally evolved versions of that community.

9.5.3   Availability of Seeds/Plants

One of the perennial frustrations of trying to use exclusively native species, even at a small landscape scale, has been their lack of availability from commercial sources. Historically, growers have produced a "standardized" list of plants, many of which are exotics, cultivars, or hybrids. The reasons for this are many and varied, but probably include the following: (1) a pioneer attitude that suggests that the native "common" species need to be improved upon or replaced by plants from another continent as symbols of "progress;" (2) selection of species that are easily propagated and/or transplanted, due to their broad amplitude of environmental tolerance; and (3) a limited knowledge of native species in any given area by the professionals (e.g., landscape architects, agronomists, and nurserymen) who influence plant production and use.

There have been some notably positive changes in this, particularly in the last 10 to 15 years, and there now is a growing number of nurseries and seed suppliers that produce some or exclusively native species. Among these, there are many reputable organizations that propagate and sell nursery-grown stock, but there are others that collect from natural areas, thereby potentially damaging those sites and decimating natural populations. Additionally, there are commercial firms that produce "wildflower" seed mixes that are implied to be native, but that may include species that are either not adapted to some of the regions in which they are sold, or that are exotic and aggressive species, such as *Daucus carota* (Queen Anne's Lace) and *Lythrum salicaria* (Purple Loosestrife).

Added to concern over some of these practices is the controversy among restorationists and ecologists as to the importance of using locally or regionally native ecotypes of native species for restorations. If one adopts the position that the propagules for a restoration must come from a relatively small radius around the restoration site, an already limited supply is limited even more.

Hence, plant availability or nonavailability becomes an important consideration in planning a restoration strategy. One response to such limitations may be to initially establish small propagtion beds, which become the "nursery" that produces plants and seeds for the rest of the site. Or, strips of native perennial plants may be established between other strips planted to easily obtained nonnative annual or short-lived perennial species, which are known to permit invasion by the desired native species.

9.5.4 Economic and Time Constraints

Somewhat related to the consideration of the availability of appropriate plants/ seeds in the desired quantity, are the combined factors of time and money in establishing a restoration strategy. For example, a restoration of a prairie using young plants is faster than establishing one by seeding, but the cost is many times higher. Except in small demonstration gardens, the cost of using started plants may be almost prohibitive. But in the restoration of a forest community, it may be necessary to plant seedling and sapling-sized trees, rather than seeds, to accelerate the development of a forest appearance and environment. In some situations, there may be reason to use a temporary planting of inexpensive, easily established but short-lived species for initial cover, with seeds of later-successional species planted either simultaneously or at a later date. Following this strategy, there might be a conscious plan for removal of the initial cover crop at some specified point.

9.5.5 Subsequent Management Prospects

Any restoration is almost certain to require conscious vegetation management practices, especially during the initial stages, for example, to suppress the development of residual exotic species that may persist or reappear. But where it is known in advance that management resources will be minimal, the restoration strategy can reflect this. For example, a prairie restoration for which periodic burning is the easiest and least expensive management technique, can be planned to include "natural" fire breaks along paths or roads, to minimize the effort required in controlling the prescribed burns. A forest restoration strategy that closely emulates a natural oldfield successional composition in its early or intermediate stage, will probably require less manipulation than one that attempts to mimic a later successional stage or climax forest composition at the outset.

9.6   Aesthetic Considerations in Restoration Design

As suggested earlier in this chapter, the restoration of aesthetic quality to a degraded landscape may be a central objective in a restoration, or it may be secondary. In any case, one of the benefits of many such projects is often an improvement in the visual quality of the site that undergoes a restoration process.

One could hypothesize that, if species selection and placement are ecologically correct, aesthetic quality will follow. However, even among different stands of naturally evolved plant community types, most observers would agree that there are variations in aesthetic quality. For example, we prefer some oak-hickory forest stands over some others, often for reasons that are related to their aesthetic characteristics.

For restorationists, it may be useful to be aware of (1) aesthetic characteristics that seem to be common to many relatively undisturbed landscapes; (2) aesthetic characteristics that are unique to particular plant community types, which may be emulated in restorations of the same communities; and (3) specific features within these stands that make them particularly memorable stands within a community type. Following is a closer look at these three levels of aesthetic consideration.

9.6.1   Aesthetic Characteristics that Are Common to Many Natural Landscapes

Dasmann (1966), in a paper entitled "Aesthetics of the Natural Environment," cites four characteristics that are typically present in relatively undisturbed landscapes, and that contribute to our perception of beauty in them: *order, diversity, health,* and *function.*

Order results from the fact that within a given environment, only a limited number of species constitute the dominant, most visible vegetation. This leads to a sense of unity or visual harmony because of the relatedness of line, form, color, and texture. In the tall-grass prairies of Iowa, Nebraska, and Kansas, for example, Weaver (1968) found that 95% or more of the vegetational cover was comprised of grasses, and of these, the vast majority of the grasses were of only 10 species. These provide the unifying background for the contrasting broad-leafed, conspicuously flowering forbs that occur in that community.

In most naturally occurring plant communities, though, the order or unity that occurs as a result of species repetition is not carried to the point of monotony. In the tall-grass prairie example, for instance, Betz and Lamp (1973) found a species density of 40 to 70 species per acre, indicating relatively small numbers of individuals providing color and textural contrast among the unity of green colors and fine textures of the grasses.

The delicate balance between order and diversity in naturally evolved communities is perhaps the characteristic which is most often missing in restorations.

Again using the prairie example, restorationists, in an effort to have a "showy" planting, tend to overemphasize the diversity, and lose the unity of the natural model.

Health, according to Dasmann, is a third component of natural beauty, and it in turn is related to the fourth component, *function.* Thriving vegetation which is flowering and producing seed is evidence of health, as is the successional progression of vegetation on a site. Function is apparent when that vegetation provides food and cover for wildlife, and when it holds soil in place and contributes to maintaining clean water in adjacent streams (Dasmann 1966).

Just as these four components contribute to beauty in natural environments, their presence is also a goal of landscape restorations.

9.6.2   Aesthetic Characteristics of Specific Plant Community Types

The next level of aesthetic consideration in restoration design is the determination of the unique aesthetic characteristics which are typical of a particular community type. This suggests identifying the essential line, form, color, and texture of the community, at each of its layers in a multilayered community, as well as its prevailing spatial structure and vegetational patterns, if they can be discerned. Such characteristics are best identified through observing a number of stands of the community type and watching for recurrent elements.

In the process of identifying these there may be particular plant species that are determined to be of great importance because of their unique aesthetic contribution. These are referred to as "visual essence" species (Morrison and Howell 1983), because of their importance to the aesthetic quality of a community. The dominant species often fall into this category, simply because of their size and numerical importance. In other cases, a species may be of greater visual importance than its size or numbers might suggest. Examples of this include plants that provide distinctive contrast with the dominant or prevalent species of the community, through their flower color or bark texture. Or they may occur in relatively small numbers, but are unique to the particular community, and thereby are "indicators" of it. Or they may provide an important, if ephemeral, flower or foliage color, or even an aroma that contributes to the aesthetic images of the community. In any case, such "essential" species are prime candidates for inclusion in a restoration.

As an example of the typical aesthetic characteristics of the tall-grass prairie, the following elements might be considered central:

1. Line/texture/movement. Because of the abundance of grasses, the prevalent lines seen in the vegetation are vertical to arching. The textural character is fine, and a unique ephemeral quality is the wavelike movement of those grasses.
2. Color. The network of grasses provide a background ranging from bright greens in the spring to various yellowish, to bluish-greens in summer, to harmonious tan, bronze, and purplish tones in the fall and winter. Within this framework, color "flecks" are provided by the forbs, generally in pastel pinks, blues, yellows, and whites in the spring to more dramatic golds and

purples later in the season. The colors of forb flowers are often partially screened or filtered by the leaves and stems of grasses.
3. Patterns. Visually apparent "drifts" of colors and textures are created by species composition variations related to microenvironmental differences and/or the plants' reproduction strategies.
4. Edges. Where prairies meet forest or shrub communities, irregular edges are typical with rounding forms of shrubs and/or small trees invading into the prairie, and prairie openings penetrating the tree and shrub zone.

9.6.3 Characteristics of Particularly Attractive Stands of a Community

Finally, at the individual stand level, there may be unique features that make some sites stand out as being particularly attractive examples of a community type. These features may be nonvegetational elements or they may be vegetational. They contribute "vividness" or "memorability" to a site (Jones 1979), in that they create a strong mental image of the place.

Among nonvegetational elements which can contribute to memorability are ephemeral phenomena such as sky and lighting conditions, or the presence of fog, mist, or snow. Other contributing factors may be: distinctive topography, geologic formations, or a view of water in a stream, lake, or waterfall.

Among vegetational elements that may contribute to memorability are plants that contrast sharply with the adjacent vegetation (e.g., the broad, coarse-textured leaves of *Silphium terebinthinaceum* among grasses in a prairie), or, simply, plants with colorful, showy flowers. Plants that are thought of as rare or unique, for example, orchids and ferns, add the quality of memorability; plants that form distinctive patterns may provide it.

The plant species that contribute memorability to a site may be the "visual essence" species for the community type, or they may be others which, by chance, are present in particular stands of that community. In either case, they would be "legitimate" species to include in a restoration in the same region with similar environmental conditions.

With the aesthethic information/observations obtained from naturally evolved landscapes, the designer/restoration planner is better equipped to plan a restoration which is simultaneously ecologically suitable and aesthetically attractive. The two goals are obviously compatible in natural environments and logically are similarly compatible in a restoration. Even stylized, somewhat simplified restoration plantings can benefit from an understanding of these elements; in fact, that understanding may be especially important in such cases, in order to evoke the aesthetic essence of the natural model.

9.7   Summary and Future Prospects

Landscape restoration as a response to previous disturbance has the potential of restoring some of the species diversity and aesthetic quality of sites that have been degraded. Over time, such restorations have the potential of becoming more like the natural models under certain conditions.

Landscape or plant community restoration planning is influenced by a number of factors: vegetation history and existing conditions on the restoration site; specific goals and objectives for the site; availability or nonavailability of seed or plants of native species; and economic factors and management prospects for the restoration.

The most reliable sources of information on appropriate species composition and community structure are relatively undisturbed natural communities themselves. Additionally, the important aesthetic characteristics of the community type can be identified for "translation" in the restoration plan.

Just as natural plant communities are dynamic and ever changing, so restorations may evolve and change over time. The vegetation history of the site, the initial species that are planted, and the establishment and management techniques that are employed will all influence the rate and direction of change in a restoration.

One of the biggest problems facing restorationists is the uncertainty of being able to obtain sufficient quantities of locally adapted propagules for native species. Development of adequate supplies, in an ethical manner, is necessary if viable restorations are to flourish.

Another challenge is in the education of professionals who are fully prepared to plan restorations. Ideally, the education for this would include elements of the education for botanists, wildlife biologists, soil scientists, horticulturists, agronomists, and landscape architects, to name a few. Within some universities, it might be possible to develop a cross disciplinary curriculum for landscape restoration which would incorporate content from each of these areas, along with a component of philosophy and ethics. But whether professionals come from the "ideal" multidisciplinary background or from just one of these areas, one requirement should be a strong community ecology field course. Closely associated with education for restoration is restoration research, which includes a concerted effort to monitor and document the evolution of sites which have undergone some level of restoration, as well as any new restorations.

Restoration, when properly planned and executed, has great potential for restoring diversity in situations where human-induced disturbances have diminished the natural diversity of a site. More generally, a productive, healthy state that is aesthetically attractive, may be facilitated by conscious restoration activity. Berger (1985; p. 5) provides this view:

> In restoration, not only are abuses halted, but the resource itself is physically repaired, and, if necessary, its missing components are replaced. Native seed varieties are sown; plants are used to bandage and rebuild the land. Roots, soil, mosses and fungi all reform a dense living mat of soil that nourishes plants, retains moisture, cleanses water percolating into the ground. Barrens become productive again. Endangered animal populations gradually revive. While certainly no longer in a pristine state, the restored resource becomes healthy, life supporting, and pleasing to the eye.

Landscape architects, with their design and planning skills, can be instrumental in facilitating landscape restorations that result in such benefits, if they

work from a solid base of ecological understanding, and/or work in conjunction
with other individuals who can provide such information and insights.

References

Berger, J. 1985. *Restoring the Earth.* Knopf, New York.
Betz, R.F., Lamp, H.F. 1973. The species composition of old settler cemetery prairies
   in northern Illinois and northwestern Indiana. *In* L.C. Hulbert (ed.), *Proceedings,
   Third Midwest Prairie Conference.* Kansas State University, Manhattan, KS.
Bradshaw, A.D., Chadwick, M.J. 1980. *The Restoration of Land: The Ecology and
   Reclamation of Derelict and Degraded Land.* Oxford: Blackwell, Oxford.
Cairns, J., Jr. 1981. Restoration and management: An ecologist's perspective. *Restor.
   Mgmt. Notes* 1(1):6–8.
Chadwick, M.J. 1982. Long-term problems of the restoration of derelict land. *In Per-
   spectives in Landscape Ecology.* Centre for Publishing and Documentation, Wag-
   eningen.
Clements, F.E. 1916. *Plant Succession: An Analysis of the Development of Vegetation.*
   Carnegie Inst. Publ. 242.
Connell, J.H., Slatyer, R.O. 1977. Mechanisms of succession in natural communities
   and their role in community stability and organization. *Am. Naturalist* 3:1119–1144.
Curtis, J.T., Greene, H.C. 1953. The re-establishment of prairie in the University of
   Wisconsin Arboretum. *Wildflower* 29:77–88.
Curtis, J.T. 1959. *The Vegetation of Wisconsin.* University of Wisconsin Press, Madison,
   WI.
Dasmann, R. 1966. *Aesthetics of the Natural Environment.* The Conservation Foun-
   dation, Washington, D.C.
Diekelmann, J., Schuster, R. 1982. *Natural Landscaping.* McGraw-Hill, New York.
Eaton, L.K. 1964. *Landscape Artist in America: The Life and Work of Jens Jensen.*
   The University of Chicago Press, Chicago.
Hutnik, R.J., Davis, G. (eds.) 1973. *Ecology and Reclamation of Devastated Lands.*
   Gordon and Breech, New York.
Jensen, J. 1939. *Siftings, and Collected Writings.* Ralph Seymour Fletcher, Chicago,
   IL.
Jones, G.R. 1979. Landscape assessment: Where logic and feelings meet. *Landscape
   Architecture* 68(2):113–121.
Morrison, D.G. 1981. Use of prairie vegetation on disturbed sites. *In Landscape and
   Environmental Design* (Transportation Research Record 822). Transportation Re-
   search Board Washington, D.C.
Morrison, D.G., Howell, E.A. 1983. Field study of native plant communities: An in-
   tensive summer course. *In Proceedings, The Landscape: Critical Issues and Re-
   sources,* 1983 Conference of Educators in Landscape Architecture. Utah State Uni-
   versity Logan.
Roberts, E.A., Rehmann, E. 1929. *American Plants for American Gardens.* MacMillan,
   New York.
Schaller, F.W., Sutton, P. (eds.) 1979. *Reclamation of Drastically Disturbed Lands.*
   American Society of Agronomy, Madison, WI.
Thorhaug, A. 1980. Recovery patterns of restored major plant communities of the United
   States: High to low altitude, desert to marine. *In* J. Cairns (ed.), *The Recovery Process
   in Damaged Ecosystems.* Ann Arbor Science Publishers, Ann Arbor, MI.
Weaver, J.E. 1968. *Prairie Plants and Their Environment.* University of Nebraska Press,
   Lincoln, NE.
Wells, B.W. 1932. *The Natural Gardens of North Carolina.* University of North Carolina
   Press, Chapel Hill, NC.

10. Patch-Within-Patch Restoration of Man-Modified Landscapes Within Texas State Parks

Tom D. Hayes,[1] David H. Riskind, and W. Lynn Pace, III

10.1 Introduction

Natural reserves serve many important purposes in anthropogenically dominated landscapes. Management of reserves in the face of past or present landscape disturbance remains a challenge. The initial interest regarding ecosystem reserves focuses on the theory and practice of preserve design and various priority schemes for preserve acquisition (May 1975; Williams 1975; Diamond 1975; Whitcomb et al. 1976; Simberloff and Abele 1976; Pickett and Thompson 1978; Polunin and Eidsvik 1979; Margules et al. 1982). Discussions deal primarily with preserve effectiveness in terms of comparisons of immigration and extinction rates developed through modeling based on island biogeographic theory. The advantages and disadvantages of large versus small preserves have been debated with biological diversity or species richness being the *raison d'etre* for preservation (Higgs and Usher 1980; Janzen 1983). Such esoterica as integration of global warming factors into preserve design have recently appeared in the ecological literature (Peters and Darling 1985).

Meanwhile, natural resource managers have been forced to take action. Most managers feared that without direct action the natural integrity of preserve islands would be eroded beyond recognition by anthropogenic landscape mod-

---

[1] Presently with: Environmental Assessment Branch, Resource Protection Division, Texas Parks and Wildlife Department, Austin, Texas.

ification. Many managers, concerned with game or endangered species, embarked on a reductionist course. More holistic managers followed the long tradition of community-oriented conservation. A major convergence of these two approaches to preserve management occurred in Leopold et al. (1963), who recognized the need for management of large herbivore herds and their predators in national parks within the context of the entire ecosystem. This ecosystem approach is now widely practiced in large resource parks and wilderness areas which, in many instances, retain the luxury of allowing natural processes already in place to merely proceed. The major management problems for such large preserves are either socioeconomic or, as in the case of acid deposition, sociopolitical.

The greater challenge to resource managers, however, is the long-term preservation of natural biota and ecosystem functions within smaller parks and preserves. Not long ago, fee simple acquisition was thought to be sufficient to effect preservation of these smaller conservation lands. Consequently, herbivore populations exploded, wetland savannas succeeded to shrub thickets, prairies soon were dominated by saplings, exotic plants grew rampantly, and some old-growth forests in which wild fires were suppressed experienced devastating crown fires. In concert with the emergence of the field of landscape ecology approximately 15 years ago, came the belated realization that many smaller conservation lands did not represent pristine landscapes but human-disturbed areas requiring relatively more stewardship and habitat restoration than larger conservation lands.

Natural resource management within disjunct parklands of limited areal extent may be compared with a network of threads interconnecting dissimilar, widely scattered fragments of a huge tattered tapestry. Due to the almost complete loss of natural integrity in the intervening landscape, management of the few remaining fragments represented by Texas state parks must counter both the local extirpation of species and processes characteristic of the natural communities and the influx of exotic and disclimax species and anthropogenic impacts. The community-centered management of parklands is often diametric to the management of surrounding lands. For example, prairie restorations use indigenous native species, although adjacent landowners plant exotic, often highly invasive grasses for "tame" pastures. Similarly, domestic herbivores are usually excluded from parklands, although they are often overstocked on surrounding lands. On the Edwards Plateau of central Texas, exotic African ungulates are excluded or removed, and overabundant white-tailed deer (*Odocoileus virginianus*) are reduced in parklands, although these species are promoted on local ranchlands. The situation is reversed in other areas, such as east Texas, where widespread poaching and habitat alteration on surrounding lands makes the reintroduction of native deer and turkey stocks largely futile in state parks.

The management approach in Texas state parks integrates a diversity of techniques from range, wildlife, forestry, and endangered species management, but the ecosystem approach is pervasive. Founded on a conservational rationale, the approach is decidedly hands-on, heuristic, and crisis-oriented (Soulé 1985). The management of smaller parks is also puristic in that well-founded ecological precepts are eagerly assimilated. However, a large amount of subjectivity is also injected due to the absence of data on many vanquished natural communities. Natural disturbance and recovery processes are mimicked to the extent possible, given our human inadequacies. As demonstrated for old-growth forest islands within human-dominated landscapes in both the northwestern (Harris 1984) and the eastern (Burgess and Sharpe 1981) United States, a "continental" source of species does not remain to replenish many species lost to local extirpations within widely scattered parklands. The system of parks itself is, in large part, the only species pool that remains. Park management must, therefore, include the active propagation of indigenous native species in order to introduce genetically accurate facsimiles of natural communities which have been eliminated or altered within specific parks. Also, accurate descriptive information about natural community compositions, ecosystem functions, and disturbance regimes must be assembled in a retrievable format for reintroduction of natural landscapes into specific parklands.

Park acquisition is controlled by economics and existing property lines and, thus, park boundaries seldom circumscribe ecological entities conducive to management such as entire watersheds. Because management strives to make each park representative of the natural communities of which it was an integral part, landscape heterogeneity is increased in order to enhance aspects of community structure and dynamics which, due to size and configurational limitations, might not occur within the park. Such restorative and manipulative techniques to promote natural community heterogeneity in disjunct parklands may be characterized as patch-within-patch management.

10.2 The Texas Landscape

10.2.1 Texas

Texas encompasses a broad range of elevations, soils, and climates within an area of 69 million hectares. Elevations range from sea level along the Gulf Coast to over 2665 m in the Guadalupe Mountains of west Texas. Some 1000 soil series encompassing eight of the nine soil orders occurring in the continental United States are found in Texas. Climates vary from subtropical near Brownsville to cold temperate in the Panhandle, with average annual precipitation ranging from more than 1375 mm in Newton County in southeast Texas to less than 200 mm in El Paso County in extreme west Texas (Kingston 1985).

The environmental complexity across the state leads to a diversity of vegetation types. East to west across Texas the vegetation changes from forests, swamps, and coastal marshes to open woodlands, prairies, and plains, and finally, in the Trans-Pecos region, to deserts interspersed with mixed conifer forests and montane meadows. Qualitative studies of the vegetation types of Texas are numerous beginning with Bray (1906) and most recently synthesized by Diamond and Riskind (1987). However, quantitative ecological studies are, in general, sparse. This lack of detailed documentation of Texas' biotic heritage r understanding of natural plant communities within the

10.2.2 Anthropogenic Disturbance

꜀ape modifications had occurred in Texas by the early 1700s, par-
n the South Texas Plains and the Texas Gulf Coastal Prairie. Vast
ı. semiwild cattle and sheep associated with Spanish colonization initiated
this ı. .dscape modification. The 20th Century, particularly the latter half, has
been a period of rapid industrialization, urbanization, and agricultural expansion
in Texas, with attendant large-scale ecosystem disruption and displacement of
native species by exotic biota.

To categorize and map existing vegetation at scales useful in statewide and
regional resource planning, the Wildlife Division of the Texas Parks and Wildlife
Department (TPWD) used ground-truthing procedures and recent LANDSAT
data and computer classification analyses (Anderson 1977; TPWD 1976, 1978,
1980). Only vegetation types which were distinguishable by physiognomy (canopy height and cover) and floristic composition at the time of the survey were
considered. Vegetation types were defined as associations of two or three dominant species and listed according to physiognomic criteria (Haas and McGuire
1974; Frentress and Frye 1975; Table 10.1). A single LANDSAT scene was
used for any given land area. Using data acquired from overflights between
1972 and 1976, classified LANDSAT scenes for the eastern two-thirds of the
state were published at a scale of 1:300,000. In the western one-third of the
state, computer classification was abandoned in favor of classifying the vegetation within land resource units previously mapped by the Bureau of Economic
Geology (BEG) of the University of Texas at Austin (Kier et al. 1977), due to
the predominance of geologic signatures in the LANDSAT data. Classification
of vegetation within BEG land resource units superimposed on LANDSAT
color-composite imagery acquired in 1979 and 1980 was accomplished at a scale
of 1:250,000 (TPWD 1981). A composite statewide map of existing vegetation
types, including cropland, urban, and water areas, was subsequently published
at a scale of 1:1,000,000 from an assembled mosaic of photographic reductions
of the smaller-scale maps (McMahan et al. 1984).

The extensive fragmentation of natural vegetation and its conversion to earlier
seral stages are apparent after examination of these statewide mapping efforts.
Clark (1985) used LANDSAT digital data and color infrared photography to
study rates of land use change for 12 areas in the eastern half of Texas (Fig.
10.1). The land use classes include woody vegetation cover, urban development
and industrial growth, and pine monoculture (Table 10.2). Natural vegetation
is being converted to urban and industrial uses at annual rates from +2.6% in
the Fort Worth-Dallas area to 11.6% in the San Antonio area. The rapid in-
discriminate loss of natural vegetation types near large cities underscores the
necessity of including natural area preservation in the urban planning process.
Clark's (1985) study also revealed rates of change of woody cover within the
study areas to vary according to short-term economic cycles associated with
timber harvesting and land clearing for pastures and croplands. Superimposed

Table 10.1. Listing of Major Physiognomic Classes Used to Standardize
Nomenclature During Mapping of Texas' Existing Vegetation[a]

Class	Definition
Grassland	Herbs (grasses, forbs, and grasslike plants) dominant; 10% or less woody canopy cover
Shrub	Individual woody plants generally less than 2.7-m tall scattered throughout arid or semiarid regions (less than 30% woody canopy cover)
Brush	Woody plants mostly less than 2.7-m tall dominant and growing as closely spaced individuals, clusters, or closed canopied stands (greater than 10% woody canopy cover)
Parks	Woody plants mostly equal to or greater than 2.7-m tall dominant and growing as clusters or as scattered individuals within continuous grasses or forbs (11–70% woody canopy cover)
Woods	Woody plants mostly 2.7–9.1-m tall with 71–100% canopy cover; midstory usually lacking
Forest	Deciduous or evergreen trees dominant; mostly greater than 9.1-m tall with 71–100% canopy cover; midstory apparent except in managed monoculture
Young forest	Various combinations and age classes of pine and hardwood regrowth resulting from the recent harvest of pine or mixed hardwood and pine forests
Marsh	Emergent herbaceous plants dominant in constantly or periodically inundated areas; 10% or less woody canopy cover
Swamp	Deciduous or evergreen trees and shrubs in constantly or periodically inundated areas; woody canopy cover greater than 10%
Crops	Includes cultivated cover or row crops used for the purpose of producing food or fiber for either humans or domestic animals
Barrier island	Smooth sloping accumulations of sand, shell, and gravel along sea and bay shores; periodically exposed unvegetated or sparsely vegetated wetlands and sand dunes

[a] Adapted from McMahan et al. 1984.

upon this cyclic pattern is a steady decrease in the overall amount of woody cover and a rapid increase in the percentage of successional shrubland in the woody cover that remains.

10.2.3 Texas State Parks

The Texas State Park System, consisting of 124[1] widely scattered units totaling 83,568 hectares, reflects the statewide diversity in physical attributes and plant communities. Ninety-nine of these units are considered natural resource parks,

---

[1] *Proof Note:* TPWD has recently acquired two additional natural resource parks totaling 2121 hectares in map area 5c (Fig. 10.2), which incorporates an important disjunct community type (e.g., pinyon-juniper) previously unrepresented in the Texas State Park System.

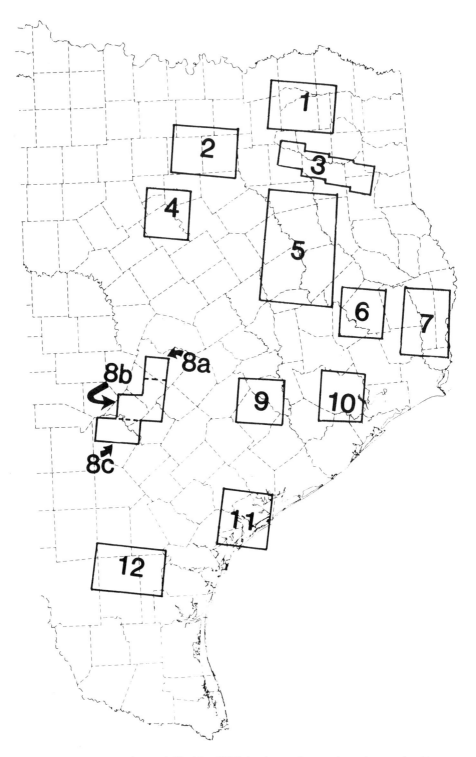

Figure 10.1. Locations of Clark's (1985) land use change detection study sites.

Table 10.2. Land Use Change Rates in Selected Areas of Texas[a]

Coverage Areas[b]	Predominant Ecologic Area(s)[c]	Land Use Class Analyzed	Period of Data Analysis	Average Net Change (%)[d]	Average Annual Change (%)[d]
Austin (8a)	Edwards Plateau, Blackland Prairies	Urban/industrial	1973–1979	+44.0	+ 7.4
San Antonio (8c)	Edwards Plateau	Urban/industrial	1973–1979	+69.6	+11.6
Ft. Worth-Dallas (2)	Cross Timbers and Prairies, Blackland Prairies	Urban/industrial	1973–1981	+20.8	+ 2.6
Houston (10)	Pineywoods, Gulf Prairies and Marshes	Urban/industrial	1972–1979	+49.0	+ 7.1
Middle Sulphur River (1)	Blackland Prairies	Woody cover	1973–1981	− 8.7	− 1.1
Upper Sabine River (3)	Blackland Prairies, Post Oak Savanna, Pineywoods	Woody cover	1973–1981	− 2.8	− 0.35
Trinity and Neches Rivers (5)	Post Oak Savanna, Pineywoods	Woody cover	1973–1981	+19.3	+ 2.4
Columbus (9)	Blackland Prairies, Post Oak Savanna, Gulf Prairies and Marshes	Woody cover	1972–1981	−12.0	− 1.3
Port Lavaca (11)	Gulf Prairies and Marshes	Woody cover	1974–1979	− 3.8	− 0.8
South Texas (12)	South Texas Plains	Woody cover	1973–1983	−21.0	−2.1
North-Central Texas (4)	Cross Timbers and Prairies	Woody cover	1972–1981	−27.9	−3.1
Lake Livingston (6)	Pineywoods	Pine monoculture	1973–1983	+6.1	+0.6
Southeast Texas (7)	Pineywoods	Pine monoculture	1974–1983	+3.0	+0.3

[a] Modified from Clark 1985.
[b] Numbers in parentheses refer to study site locations depicted in Figure 10.1.
[c] According to Gould et al. 1960.
[d] Quantitative change in number of pixels assigned to land use class between the two sets (earlier and later years).

although the size and integrity of protected natural communities varies from park to park (Fig. 10.2; Table 10.3). Ranching, farming, forestry, urbanization, hydrologic alterations, energy exploitation, and introductions of exotic plants and animals have dramatically altered the composition and structure of natural communities within many state parks. In response to growing awareness of such widespread landscape alterations and other environmental concerns, the Parks and Wildlife Commission established the Resource Management Section (RMS) within the Parks Division of the TPWD in 1972. The RMS was given the task of providing integrated natural resource management to balance recreational use and, in some instances, continued energy exploitation with the maintenance or re-establishment of natural community integrity and processes (Riskind 1977). After four decades of intensive public use, older parks had developed critical resource problems, including degraded natural communities and soil compaction with resultant erosion. Modifications of natural hydrologic regimes had also impacted many parks, especially in coastal areas.

Most resource management in Texas state parks is directed toward the maintenance or restoration of natural plant communities, although the reintroduction of vertebrate fauna often occurs as a final step in the restorative process. Through selective brush control, direct seeding of native grasses and forbs, native tree and shrub plantings, and other measures, community restoration is achieved sooner than if left to natural processes. Although management within Texas state parks emphasizes the rapidly declining old-growth communities, many successional communities are also significant and require maintenance. For example, the pine-heath fire subclimax community within two parks in Bastrop County is significant due to being the southwesternmost outlier of its community type and supporting several endangered species. Thus, the community is periodically prescription burned to prevent succession to a hardwood overstory. The emphasis is on pragmatism, for just as individual parks exist as remnants in an environmental matrix, they also exist within a sociopolitical matrix often emphasizing an extractive land ethic. Resource-based nonconsumptive recreation within the parks also requires a pragmatic approach to the location and maintenance of use areas, based on Stankey's (1985) tenet regarding the limits of acceptable change.

In most community restorations, native plants are used near their point of origin and within their area of biogeographic distribution. The procurement of appropriate plant materials is greatly assisted by the RMS's close relationship with the USDA Soil Conservation Service's (SCS) Plant Materials Center (PMC) at Knox City, Texas (Pace et al. 1975). Since the Knox City PMC's establishment in 1964, 29 hectares of irrigated cropland have been devoted to the selection, development, and production of plants which are needed for various conservation and restoration projects throughout Texas and western Oklahoma (Fig. 10.3). Due to past evaluations, a number of native species have been made available to commercial seed producers and are now available for public use. In the early 1970s, a cooperative agreement between the SCS and the TPWD provided for a biologist with the RMS to be headquartered at the PMC to train and assist in all aspects of the PMC's operation, in addition to assisting area

1. Pineywoods
  1a.   Oak-Hickory-Pine Forest
  1b.   Southern Mixed Forest
  1c.   Southern Floodplain Forest (not mapped)
2. Central Texas
  2a.   Oak Woodland
  2b.   Eastern Cross Timbers
  2c.   Western Cross Timbers
  2d.   Bottomland Forests (not mapped)
  2e.   Fayette Prairie
  2f.   Blackland Prairies
  2g.   Grand Prairie
  2h.   North Central Prairie
3. Gulf Coast
  3a.   Outer Barrier
  3b.   Estuaries (not mapped)
  3c.   Salt Meadows and Marshes
  3d.   Inner Barriers and Salt Domes (not mapped)
  3e.   Coastal Prairie
  3f.   Bottomland Forests (not mapped)

4. South Texas Plains
  4a.   Coastal Sand Plains
  4b.   Subtropical Zone
  4c.   Bottomland Forests (not mapped)
  4d.   Brush Country
5. Edwards Plateau
  5a.   Lampasas Cut Plain
  5b.   Llano Area
  5c.   Balcones Canyonlands (subregion shown
        is a mixture of canyonlands with Oak-
        Juniper Savanna on the uplands)
  5d.   Stockton Plateau
  5e.   Mesquite Belt
  5f.   Oak-Juniper Savanna
6. Rolling Plains
  6a.   Escarpment Breaks
  6b.   Canadian Breaks
  6c.   Mesquite Plains
7. High Plains (not divided into subregions)
8. Trans-Pecos
  8a.   Mountain Ranges
  8b.   Desert Grasslands
  8c.   Desert Scrub
  8d.   Howard Bolson and Salt Flats
  8e.   Monahans Sand Hills

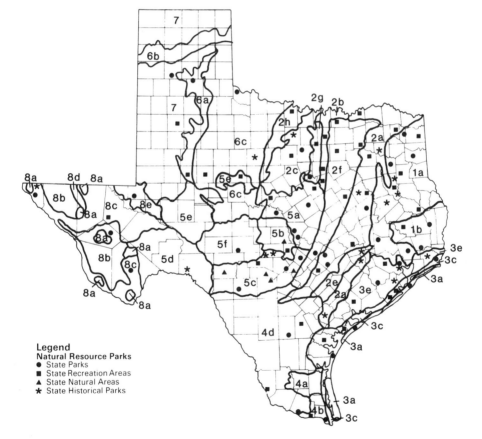

Legend
Natural Resource Parks
● State Parks
■ State Recreation Areas
▲ State Natural Areas
★ State Historical Parks

Figure 10.2. Distribution of 99 natural resource parks within Texas State Park System relative to natural regions of Texas.

Table 10.3. Size Distribution of 99 Natural Resource Parks Within Texas State
Park System[a]

	Number of Parks	Average Size (hectare)
Administrative category		
State parks	38	1402.2
Natural areas	5	987.1
Recreation areas	39	297.6
Historical parks	17	177.1
All categories	99	735.7
Size distribution (hectare)		
Less than 100	14	
100–500	49	
500–1000	19	
1000–3000	13	
3000–5000	0	
Greater than 5000[b]	4	

[a] Of 124 units in the system, 25 consist of historic structures and sites and fishing piers which average only 8 hectares in size and do not contain significant natural communities. The 13,223-hectare Choke Canyon park site, McMullen and Live Oak counties, includes a 10,530-hectare reservoir, so an areal extent of 2693 hectares for this park is used in this tabulation.
[b] The natural resource park presently containing the largest land area in the system is 6643-hectare Palo Duro Canyon State Park, Randall County. When ongoing acquisition is complete, the Franklin Mountains park site, El Paso County, will encompass virtually the entire desert mountain range and total approximately 14,175 hectares.

parks in the application of plant materials and other resource management. In exchange, the center annually allocates a portion of its materials to the TPWD, including materials grown for specific parks. Though the allocated quantities are less than bought commercially, the Knox City PMC's stocks are essential for the inclusion of locally adapted materials not commercially available. The other 23 PMCs throughout the United States represent similar opportunities for restoration. The RMS has also cooperated with plant propagation programs at universities and wholesale nurseries to obtain indigenous materials.

10.3   Management Examples

10.3.1   Prairie Restoration in the Rolling Plains

During the past 12 years, more than 405 hectares of abandoned cropland and depleted rangeland within state parks in the Rolling Plains of north-central Texas have been restored to native prairie by direct seeding. Potential restoration sites generally lack fertility and are dominated by weedy vegetation. Coupled with weather characterized by periods of drought, temperature extremes, and strong winds, seeding operations are tempered by less-than-optimum growing conditions. The rate of restoration was estimated during a 49-hectare prairie restoration at Copper Breaks State Park in southern Hardeman County. In-

Figure 10.3. Aerial of Knox City Plant Materials Center. (Photo courtesy of W.L. Pace.)

tensive seedbed preparation, involving successive disc-plowing and cover/litter crop establishment, destroyed the existing weedy vegetation. Four seed mixtures composed of different proportions of 15 native species were seeded with a range drill in two range sites in March 1974. Data obtained from permanent transects during each of the three years after seeding indicated a sharp increase in relative importance and cover of seeded species in the third year, a pattern which has been repeated in subsequent prairie restorations in the Rolling Plains.

Important to all prairie restorations is the use of different seed mixtures within each distinct microenvironment. This maximizes efficiency and success by anticipating patchwise differences in species composition within the original prairie. A 1977 prairie restoration in the Rolling Plains at Lake Colorado City State Recreation Area, Mitchell County, for example, used 22 accessions representing 17 species of native grasses, forbs, and a woody plant to formulate eight seed mixtures corresponding to the complex environmental matrix within an 80-hectare restoration area.

An 150-hectare prairie restoration at Caprock Canyons State Park used minimal seedbed preparation to restore cropland. Located in southeastern Briscoe County, the park encompasses 5632 hectares along the Escarpment Breaks transition between the Rolling Plains to the east and the High Plains to the west. Cattle were first pastured on what is now parkland in 1877. At the time of park acquisition in 1975, the area was terraced cropland on which wheat, grain sorghum, and cotton had been grown continuously for about 60 years. Upon initiation of the prairie restoration in 1980, the old fields supported a diverse vegetation dominated by such successional species as common sunflower *(Helianthus annuus)*, Russian thistle *(Salsola kali)*, horse-weed *(Conyza canadensis)*, and other weedy annual forbs and grasses. Perennial grasses such as silver bluestem *(Bothriochloa saccharoides* var. *torreyana)*, sand dropseed *(Sporobolus cryptandrus)*, and plains bristlegrass *(Setaria leucopila)* had also established.

After terraces were partially leveled and cross-fencing removed, the old fields were burned in March 1980. This less intensive seedbed preparation accelerated the seeding schedule by one or more years, reduced expense due to omission of discing and cover/litter crop establishment, created a clean seedbed without topsoil disturbance, temporarily removed competitive species including woody species, and reduced wind erosion due to spring regrowth of existing perennial species. In early May 1980, 1745 bulk kg of seed (913 kg pure live seed) representing 21 separate accessions and 17 native species (ten grasses, three forbs, three legumes, and one woody plant) were drilled into the burned stubble. Different seed mixtures were used on the two range sites.

Weedy species, such as Johnsongrass *(Sorghum halepense)*, threeawn *(Aristida spp.)*, sand dropseed, and several annuals, are present but decreasing in the restored prairie. The relative importance of all seeded species has steadily increased since 1980, with sideoats grama *(Bouteloua curtipendula)*, blue grama *(B. gracilis)*, little bluestem *(Schizachyrium scoparium)*, Engelmann daisy *(Engelmannia pinnatifida)*, and bush sunflower *(Simsia calva)* now being well es-

Figure 10.4. Restored native prairie, Caprock Canyons State Park. (Photo courtesy of W.L. Pace.)

tablished (Fig. 10.4). As with other prairie restorations and native rangelands, management continues. Selective brush control for species such as honey mesquite *(Prosopis glandulosa)* and Pinchot juniper *(Juniperus pinchotii)* is carried out either mechanically or chemically. Prescribed burning is also planned to reduce woody plant colonization, remove excessive litter, reduce competition from woody species, and stimulate growth of fire-adapted native species. Additional overseeding of appropriate species was begun in 1984 and continues periodically to further "nudge" the species composition in the direction of indigenous mixed-grass prairie.

Once the seeded species were established, the selective reintroduction of original faunal components of the prairie ecosystem was initiated. Black-tailed prairie dogs *(Cynomys ludovicianus)* were live-trapped from a population in Big Spring State Recreation Area (Howard County) and relocated to the Caprock Canyons restored prairie in June 1983. Of the 17 introduced, three were adults and 14 were weaned, half-grown juveniles. The high proportion of young animals provided a more adaptive population for release into the unfamiliar environs of the restoration area. The release site was located 90 m from an interpretive facility, and two methods were used to imprint the prairie dogs to this specific location. First, to simulate the closely cropped character of an established prairie dog town, two adjoining circles 30 m in diameter were shredded to a height of 5 cm before release; and second, temporary burrows within the shredded areas were prepared. Holes were dug at an angle of approximately 45° with northerly openings and an open-ended, wire-mesh cyclinder was placed in each artificial burrow. One or more animals and a small amount of freshly cut grass were placed into each cylinder before the top was sealed and the confined animals allowed to burrow (Fig. 10.5). The cylinder was removed after 24 hours. The reintroduction was successful, with the prairie dog town confined for the past three years to the prepared release site, though expansion of the town is now proceeding. As a concession to the local residents and officials who have long mandated the total eradication of prairie dogs as pests, the prairie dog population will be controlled to prevent expansion beyond the park boundary.

To further simulate natural herbivory, an 610-hectare area was fenced in 1984, including the 150-hectare restored prairie discussed above, an adjoining 170-hectare Indiangrass *(Sorghastrum nutans)*-sideoats grama restored prairie, and a 290-hectare area of gypseous badlands and riparian habitat. A pair of bison *(Bison bison)* were reintroduced into half of the restored prairie acreage. Pronghorns *(Antilocapra americana)*, for which access to both summer range (restored prairie) and winter range (badlands and riparian habitat) has been provided by visually unobstrusive, strategically placed fenced corridors, were also recently reintroduced.

10.3.2   Habitat Restoration in the Lower Rio Grande Valley

The lower delta of the Rio Grande in south Texas contains remnants of one of only two subtropical biotas of the continental United States. Delineated by Blair (1950) as the Matamoran subdistrict of the Tamaulipan biotic province, the delta contains floral and faunal taxa which occur nowhere else in the United

Figure 10.5. Imprinting of prairie dogs using temporary burrows, Caprock Canyons State Park. (Photo courtesy of W.L. Pace.)

States. Although the distinctive biota is poorly documented, a few good qualitative accounts exist (Clover 1937; Johnston 1963; Everitt and Gonzales 1983; Neck et al. 1984).

Due to its favorable soils and climate, the delta has attracted large-scale agricultural development since the early 20th Century. Today, recreation-oriented residential development is also becoming important. For example, a pending international development is to include approximately 5125 hectares in the United States and a similar area in Mexico within the lowermost portion of the delta, encompassing undeveloped wetlands, barrier islands, and *loma* thorn shrublands. Approximately 5% of native habitat remains in hundreds of small remnant tracts of thorn shrubland, thorn woodland, and riparian forest north of the U.S.-Mexico border. None of the mesquite *(Prosopis spp.)* savanna and the open subtropical grassland remains (Johnston 1963). Due to the construction of extensive levee systems to control flooding and several reservoirs and associated irrigation networks upstream on the Rio Grande and the Rio Concho, relict riparian forest patches are no longer seasonally flooded. Consequently, these patches are becoming less diverse with honey mesquite displacing mesic climax species such as Texas ebony *(Pithecellobium flexicaule)*, cedar elm *(Ulmus crassifolia)*, sugarberry *(Celtis laevigata)*, anacua *(Ehretia anacua)*, and Mexican ash *(Fraxinus berlandieriana)*.

In belated response to this continuing destruction of unique subtropical habitats, the TPWD, the U.S. Fish and Wildlife Service (FWS), and other agencies and organizations began to acquire preserves. Acquisition of parcels adjacent to the Rio Grande has been given priority to assemble corridors connecting disjunct remnants with the major biotic core area, the Gulf Coastal Plain of Tamaulipas. The acquisition in 1942 of the most significant preserve, Santa Ana National Wildlife Refuge, began a major effort by the FWS. The FWS land protection plan (FWS 1985) combines fee and easement acquisition alternatives to protect 43,538 hectares of the best remaining tracts of 10 subtropical habitats within the Lower Rio Grande Valley. However, less than 20% of the remnants identified by the FWS have thus far been acquired. Current trends suggest complete destruction of unacquired habitat remaining in private ownership within five years (FWS 1985). Due to the high cost of the remaining native habitat, current funding is insufficient to meet acquisition goals. Even if acquisition of all significant remnants was possible, the total land area within preserves might still be insufficient to protect viable populations of many biota and provide necessary buffer strips and migration corridors.

To offset the inevitable demise of many remnants, the TPWD has for 30 years restored cultivated farmland, a considerable proportion of many acquisitions, to native shrubland and woodland. For the past five years, the FWS has cooperated with the TPWD in this restoration effort. The TPWD first conducted restoration research during the late 1950s on the Longoria Unit of the Las Palomas Wildlife Management Area (Fig. 10.6). This early research emphasized five native woody species: huisache *(Acacia farnesiana)*, Texas ebony,

Figure 10.6. Aerial of the Longoria Unit of the Las Palomas Wildlife Management Area. (Photo courtesy of J.H. Dunks.)

anacua, granjeno *(Celtis pallida)*, and brasil *(Condalia hookeri)*, which provide important nesting habitat for the declining white-winged dove *(Zenaida asiatica)*. Seedlings of these species were transplanted from existing native stands into test plots. The evaluation of a variety of irrigation, fertilization, and weed control techniques determined restoration to be feasible but costly due to reliance on extensive hand labor (Waggerman 1978). Today, much more efficient techniques of brush restoration are being used. Large quantities of seed are collected and stored for germination under optimum conditions (Fulbright 1985). Four- to six-month-old greenhouse seedlings are planted at the restoration site with a chisel-type tree planter pulled by a four-wheel-drive tractor. During 1985, planting costs were reduced to only 15% of those of the previous decade, primarily due to mechanization (George 1985). Ongoing research emphasizing plant-growth hormones (gibberellic acid) and temporary shade structures on restoration plots may reduce costs even more.

These techniques are now being applied in the restoration program at the Resaca de la Palma park site in Cameron County. The 448-hectare park site contains 227 hectares of native shrubland and wetland *(resaca)* habitat, with the remainder being cropland. The basic restoration concept consists of the establishment of insular patches of native shrubland within former cropland to provide seed sources for subsequent colonization by means of natural succession. Based upon previous studies related to both the species composition and spacing of plantings (Waggerman 1978), an idealized placement has been developed as a guide for establishment of 0.4-hectare restoration patches (Fig. 10.7). Rapid canopy closure within each patch, especially at the perimeter, minimizes active control of troublesome weedy species such as Roosevelt weed *(Baccharis neglecta)*, Johnsongrass, and buffelgrass *(Cenchrus ciliaris)*. Seedling spacing of 3.7 × 4.6 m is most cost-effective, while still providing rapid canopy closure. To imitate natural successional processes, 75% of the species composition consists of huisache, a rapid-growing, early successional, small tree (Bush and Van Auken 1985). Slower-growing species, such as Texas ebony, granjeno, anacua, and coma *(Bumelia celastrina)*, are interspersed within the remainder of the patch to provide heterogeneity, especially after the relative importance of huisache begins to decline after 20 to 25 years.

At the Resaca de la Palma park site, a further increase in cost-effectiveness has been achieved by adhering to UNESCO's Man and the Biosphere Program's basic precept of integrating regional agricultural practices with habitat restoration. A farm lease agreement in 1983 specifies a modern unidirectional version of shifting cultivation. In lieu of cash lease payments, the lessee is annually required to prepare seedbeds for native brush restoration, irrigate existing restoration areas, and plant wildlife food crops. Therefore, less area is available to the lessee each year to produce cash crops as more is restored. Besides enhancing the use of restored native shrubland by targeted wildlife species such as white-winged dove, the establishment of adjacent wildlife food crops augments the animal vector for seed influx from native source pools. In this manner, the natural successional process is amplified.

```
H H H H H H H H H H H H H H H H H H H H H H H H H H H H H H H H H H H H H H H H H H H H H H H H H H
H E H H C H H G H H A H H E H H C H H G H H A H H E H H C H H G H H A H H E H H C H H G
A H H E H H C H H G H H A H H E H H C H H G H H A H H E H H C H H G H H A H H E H H C H
H E H H C H H G H H A H H E H H C H H G H H A H H E H H C H H G H H A H H E H H C H H G
A H H E H H C H H G H H A H H E H H C H H G H H A H H E H H C H H G H H A H H E H H C H
H H H H H H H H H H H H H H H H H H H H H H H H H H H H H H H H H H H H H H H H H H H H H H H H H H
```

Symbol	Species	No.
H =	Huisache =	204
E =	Ebony =	16
C =	Coma =	16
G =	Granjeno =	14
A =	Anacua =	14
	Total =	264

Spacing of Seedlings: 3.7 × 4.6 meters

Figure 10.7. Idealized placement of seedlings in a 0.4-hectare plot, Lower Rio Grande Valley reforestation project.

10.3.3 Habitat Manipulation to Promote Natural Heterogeneity

In addition to the restoration of natural communities by direct seeding and transplanting, management of smaller resource parks includes the manipulation of disturbed biotic communities to promote landscape heterogeneity. Manipulation allows natural processes to be either reintroduced or continued within park landscapes.

Since the early 1970s prescribed burning has been used as a management tool in state parks throughout Texas, with the exception of arid far west Texas and subtropical extreme south Texas. Fire is reintroduced as a natural community process primarily during the cool season to prohibit the invasion of woody species into prairies and to restore the open mosaic of oak (*Quercus spp.*) and oak-Ashe juniper (*Juniperus ashei*) woodlands on the Edwards Plateau of central Texas. Management plans for specific parks include burn-grid maps delineating different burn cycles for subareas, based on fuel load, successional stage, natural and artificial fire breaks, and topography. Only portions of individual parks are burned in any given year, both to enhance landscape heterogeneity and to not overtax escape mechanisms and seedbanks of the affected biota. In parks within the Gulf coastal prairies, a natural fire regime is allowed. The typical lightening-generated summer fire along the coast seldom exceeds 150 hectares due to environmental patchiness and is easily accommodated within large coastal parks.

In contrast to the community-oriented approach within state parks, the TPWD's Wildlife Division often utilizes prescribed burning to enhance habitat conditions for particular game species. For example, hot summer fires which ignite oak mottes and promote resprouting enhance deer production. Research at the Kerr Wildlife Management Area, Kerr County, Texas, has detailed microhabitat and interspecific temperature differences during fire in oak woodlands on the Edwards Plateau (Stone 1984; Fonteyn et al. 1984). Differences in the fire behavior of dominant species has allowed the creation of different landscapes by burning during different seasons and weather conditions (Baccus 1983, 1984; Cross 1984).

Petrochemical exploration and production, pipeline and other utility easements, grazing, and similar impacts occur in addition to park development, due to legislative mandate, deed restrictions, or unconveyed mineral rights. Such anthropogenic disturbances are manipulated to minimize impacts to natural communities and processes. Due to their importance as ecosystem linkages and migration corridors, riparian areas in parks are generally not modified. During the construction of roads and trails, culverts, water bars, and similar structures are installed to maintain natural drainage patterns. Energy exploration, such as seismic surveys, is routinely excluded from drainages and other sensitive areas. Oil and gas drilling and production facilities are preferentially located on relatively flat ridge tops, so that any sedimentation or other pollutant discharge which circumvents on-site safeguards will be more distant from streams to facilitate subsequent clean up and assimilation.

Anthropogenic disturbances may alternatively be viewed as offering manipulative opportunities for patch-within-patch management within parks. Active control of the southern pine beetle (*Dendroctonus frontalis*) as mandated by state legislation in east Texas parks, for example, is partially ameliorated by specifying minimal damage to the hardwood understory during remedial timbering. Maintenance of the hardwood understory allows the natural dominance of hardwoods to occur in these patches, which also leads to a long-term decrease in the incidence of beetle outbreaks.

Other examples of opportunities derived from anthropogenic disturbances involve the perpetuation of native successional communities. The decline of these communities is of particular concern in east Texas parklands of old-growth pine-hardwood forests surrounded by pine plantations and improved pastures. Mitigative requirements during oil and gas drilling and production in these parks include the use of native successional species supplied by the RMS during site stabilization and habitat restoration (Fig. 10.8). Therefore, a more diverse native seedbank is available for secondary succession after natural gap-phase disturbances, such as wind throw, in old-growth forest adjacent to oil and gas extraction facilities, although natural succession by woody species proceeds in the seeded areas. Similar opportunities for maintaining native successional communities occur along roadsides and other right-of-ways.

Human-modified landscapes within Texas state parks are also manipulated to increase heterogeneity in order to stabilize declining populations of certain species, as illustrated in the following two examples.

10.3.3.1 Creation of a Refugium for Endangered Desert Fishes

The Commanche Springs pupfish (*Cyprinodon elegans*) and the Pecos gambusia (*Gambusia nobilis*) are both listed as endangered by the FWS and the TPWD. The two species once thrived in stenothermal spring runs surrounded by fresh-

Figure 10.8. Native planting within buffer area surrounding oil well pad, Tyler State Park. (Photo courtesy of D.H. Riskind.)

water marshes, called *ciénegas*, in west Texas. Groundwater pumping since the late 19th Century has led to the demise of several of these springs and associated ciénegas (Hendrickson and Minkley 1985). Consequently, some populations of these two protected fish species are now extirpated (Echelle 1975), a plight increasingly common to endemic desert fishes in the western United States (Miller 1961; Hubbs and Echelle 1972).

San Solomon Springs, one of the major springs (1.14 m³/sec), lies within Balmorhea State Park, Reeves County. The springs are at a sufficiently low elevation within the aquifer so that flow continues, though in gradually decreasing amounts. Civilian Conservation Corps development in the 1940s converted the spring to a 1-hectare spring-fed swimming pool, the *Sporobolus*-dominated ciénega to a turf of Bermudagrass *(Cynodon dactylon)* interspersed with planted Rio Grande cottonwoods *(Populus wislizenii)*, and the vegetated spring run to a concrete-lined irrigation ditch. Such habitat destruction eliminated a large majority of the local Commanche Springs pupfish. The relict population continued to be threatened by the increasing appropriation of water for irrigation and the inadvertent introduction of the sheepshead minnow *(Cyprinodon variegatus)*. Although outcompeted by the sheepshead minnow in quiet pools, the Commanche Springs pupfish survived within the moderately rapid current of a 100-m stretch of irrigation ditch effectively separated by an outflow over a short drop from pool habitats downstream dominated by the sheepshead minnow.

In response to the endangered status of both fish species, the TPWD in concert with other specialists including the FWS and the Desert Fishes Council, designed and constructed in 1975 a refugium within Balmorhea State Park for the species using shunted flow from San Solomon Springs (Fig. 10.9). The refugium consists of a gunite-lined canal 300 m long, which mimics the habitat heterogeneity of the spring run-ciénega habitat of the two fish species. The Comanche Springs pupfish colonized the refugium naturally from the population upstream within the irrigation ditch, whereas the Pecos gambusia, which had been extirpated from the San Solomon Springs system, was restocked from a surviving population at nearby Griffin Springs. In adherence to the community management approach, natural predators, primarily Mexican tetra *(Astyanax mexicanus)* and green sunfish *(Lepomis cyanellus)*, are also allowed within acceptable limits.

A patchy distribution of both emergent and floating vegetation is maintained. Stonewort *(Chara spp.)*, a floating macrophyte, is allowed to form continuous mats up to 15 m in length, at which time 3-m swaths are mechanically cut at 15-m intervals along the channel. The rough texture of the channel lining is designed to allow a random distribution of small patches of emergent vegetation, including *Typha spp.*, *Scirpus spp.*, *Eleocharis spp.*, and *Cyperus spp.*, which are also subject to mechanical control. No shading by overstory vegetation is allowed, so that stenothermal conditions are maintained.

Other aspects of habitat heterogeneity that are artificially maintained relate to the substrate thickness and the depth and flow of water. To replicate poorly understood substrate requirements, silt from the irrigation canal which supported the surviving breeding population of the pupfish was used to maintain a substrate thickness of 5 to 15 cm in the refugium. Excess substrate is removed during the mechanical harvesting of aquatic plants. Water flow through the refugium is maintained by an auxiliary pump. Shallow riffle areas, which are important both for breeding and predator avoidance, alternate with deeper pools throughout the refugium.

The refugium at Balmorhea State Park is an example of reintroducing and maintaining natural heterogeneity within a human-made aquatic environment. The two endangered fish species have successfully reproduced each year since the refugium was established. The detailed management plan and monitoring system have recently been recommended by the Rio Grande Fishes Recovery Team for inclusion as a component of the Commanche Springs Pupfish Recovery Plan.

10.3.3.2   Enhancement of Avian Habitat by Promoting Woodland Heterogeneity

The golden-cheeked warbler *(Dendroica chrysoparia)* and the black-capped vireo *(Vireo atricapillus)* have fairly synchronous breeding ranges largely restricted to the Balcones Escarpment on the eastern margin of the Edwards Plateau in central Texas. Both species are declining due to habitat conversion and nest parasitism by brown-headed cowbirds *(Molothrus ater)*. Because of its rapid

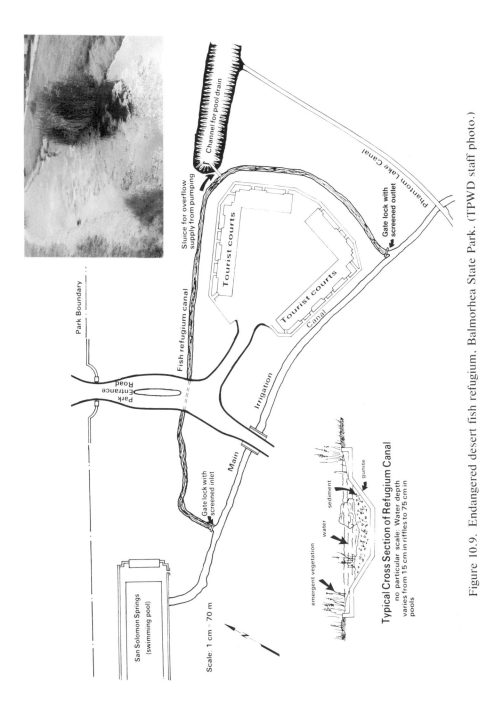

Figure 10.9. Endangered desert fish refugium, Balmorhea State Park. (TPWD staff photo.)

decline, the vireo is now being considered for listing as endangered by the FWS (Marshall et al. 1985) and the TPWD. Distribution, breeding biology, and habitat requirements for both the warbler (Pulich 1962, 1976; Kroll 1974, 1980) and the vireo (Graber 1961; Marshall et al. 1985) are relatively well known. Both are edge species inhabiting narrow ecotonal areas between oak-Ashe juniper woodlands and grasslands.

Habitat manipulation to arrest succession within shrub patches is a viable management tool for both the warbler and the vireo. Under natural conditions, these patches persist due to either edaphic factors, such as shallow soil over limestone on ridge tops or steep slopes, or cyclic disturbances, such as fire, seasonal deer browsing, and pest and disease outbreaks. Little direct management for the warbler is necessary, because the small average size (608.5 hectares) of the 12 parks in the breeding range and the development of park facilities, roads, and trails creates an abundance of edge habitat within mature Ashe juniper-hardwood woodlands. However, experimental pruning and thinning to form more open areas amenable to the warbler within dense Ashe juniper stands was undertaken in Meridian State Recreation Area in the late 1960s. Subsequent censuses indicated an increase in the number of breeding pairs of warblers at the periphery of these areas.

The management plan being implemented focuses on the distinctive habitat prerequisites of the vireo, but also benefits the warbler. Ephemeral patches composed of small trees (Ashe juniper and hardwoods) interspersed with successional shrub species, such as white shin oak *(Quercus sinuata* var. *breviloba)*, sumac *(Rhus spp.)*, and Texas persimmon *(Diospyros texana)*, are maintained by mechanical coppicing primarily using volunteer labor. In particular, shrubs 1 to 2 m tall with dense canopies extending to the ground are maintained, because they are required for vireo nesting. Another habitat requirement which is maintained is the dissection of these shrub patches by bare ground, rock, and herbaceous cover. Also, the management plan specifies a lack of active response, such as reforestation, to the ongoing epidemic of the oak wilt fungus *(Ceratocystis fagacearum)*, which is causing extensive mortality to two of the dominant overstory species, plateau live oak *(Quercus fusiformis)* and Texas oak *(Q. texana)*, in upland woodlands of the Edwards Plateau. If the resultant increase of shrubs within these overstory openings is found to be appropriate for the two species, some of these successional patches may be maintained through coppicing. In the future, small but hot fires may be used to further promote a regrowth of oak and sumac shrubs.

The habitat manipulation described is proposed for each of the 12 state parks within the synchronous breeding range of the two avian species. Management for the two species is community oriented in that numerous nontarget plant and animal species are also maintained by increased landscape heterogeneity. The ultimate objective is to concentrate the habitat requirements of the two species both spatially and temporally within the confines of each park. For these migratory species requiring small home territories, this system of small preserves scattered across the breeding range may provide greater protection than one or a few large preserves. The approach integrates the structural and compositional differences in the breeding habitat of the two species occurring across the Balconian biogeographic province represented by the parks. For example, active management will be emphasized in parks east of the 98th meridian within the two species' synchronous breeding range, due to the higher precipitation and lower evaporation rates in this area which lead to more rapid succession and overstory closure within the shrub patches.

10.4   Conclusion

The overwhelmingly private ownership of land in Texas has placed the state in the forefront of anthropogenic landscape modification and accompanying concerns. The perspective of the Texas State Park System regarding the need to restore human-modified landscapes is firmly based on the here and now and will only remain viable for perhaps 50 years hence. Presently, management of park landscapes focuses on recovery from historic disturbances tied to previous land use. As parkland restoration proceeds, the management posture will increasingly shift toward the maintenance of restored biotic communities. As conversion to anthropogenic landscapes intensifies at the periphery of the parks, concerns regarding the conservation of insular populations and the exclusion of exotic species and other human-related disturbances will take precedence. For example, the recent emphasis in Texas of acquiring parklands near major metropolitan areas will lead to many parks being engulfed, not in homogenized agricultural landscapes, but in urban landscapes teeming with a diversity of exotic and extralocal species. Will choices need to be made regarding which organisms and biotic communities can be accommodated in future resource parks? Will acquisition policy be altered to adhere to ecologically defined boundaries which are more "defensible?" Will supplemental buffer areas and connective corridors be acquired in time to protect existing preserves? Or will smaller resource parks cease to function as ecosystem preserves and, instead, become botanic and zoologic gardens?

The integrity of preserves in the Rio Grande delta has now reached this important crossroad. We have reached this point, for the most part, through the now familiar crisis management format, without benefit of rigorous discussion of preserve design or the other critical parameters so clearly discussed by Soulé and Simberloff (1986) and White (1986). The goal of the various cooperative agencies has simply been to acquire everything feasible in the delta. Only recently have concerns emerged about whether there is significant remaining habitat to truly protect species such as the ocelot *(Felis pardalis)* and jaguarundi *(F. yagouaroundi)* (Tewes and Everett 1982) or whether preserves are required along riverine corridors to ensure preserve integrity. And, while buffer areas have been acquired and in some cases restored to native shrubland, each remnant patch has uncomfortably high "edge to core" ratios and indeed is described aptly as a garrison reserve (Soulé and Simberloff 1986).

The experience with preserves in the Rio Grande delta is relevant to the whole issue of ecosystem protection, both in the sense that there is much to gain from research and application if we succeed and, if we fail, we lose a unique subtropical biota. The lesson is that ecosystem preserves must integrate smaller preserves as functional components if long-term preservation is to be achieved. Unfortunately, due to the general absence of basic research in conjunction with ongoing landscape restorations, this lesson is still largely to be learned.

Acknowledgements

The authors are grateful to David Diamond, Paul Fonteyn, Fred Gehlbach, Larry Gilbert, Monica Turner, and two anonymous reviewers for their candid comments on earlier drafts. Personnel of the TPWD who also provided helpful comments include Ron George, Mike Herring, Craig McMahan, Raymond Neck, Floyd Potter, Ron Ralph, Bernadette Rittenhouse, Gary Waggerman, and John Williams. John Emerson, also of the TPWD, was the draftsman.

References

Anderson, S.A. 1977. Pattern recognition for habitat mapping. Texas Parks and Wildlife Department, Austin, TX.

Baccus, J.T. 1983. Effects of prescribed burning upon white-tailed deer habitat. Job Number 39, Federal Aid Project Number W-109-R-6, Texas Parks and Wildlife Department, Austin, TX.

Baccus, J.T. 1984. Effects of prescribed burning upon white-tailed deer habitat. Job Number 39, Federal Aid Project Number W-109-R-7, Texas Parks and Wildlife Department, Austin, TX.

Blair, W.F. 1950. The biotic provinces of Texas. *TX. J. Sci.* 2:93–117.

Bray, W.L. 1906. Distribution and adaptations of the vegetation of Texas. Bulletin no. 82, series no. 10. University of Texas, Austin, TX.

Burgess, R.L., Sharpe, D.M. (eds.) 1981. *Forest Island Dynamics in Man-Dominated Landscapes.* Springer-Verlag, New York.

Bush, J.K., Van Auken, O.W. 1985. Light requirements of *Acacia smallii* and *Celtis laevigata* in relation to secondary succession on floodplains of south Texas. *Am. Midland Naturalist* 113:118–122.

Clark, B.V. 1985. Landuse change rates in selected areas of Texas. Remote Sensing Center, Texas A&M University, College Station, TX.

Clover, E.U. 1937. Vegetational survey of the Lower Rio Grande Valley, Texas. *Madroño* 4:41–66, 77–100.

Cross, D.C. 1984. The food habits of white-tailed deer on the Kerr Wildlife Management Area in conjunction with prescribed burning and rotational livestock grazing systems. Master's thesis, Biology Department, Southwest Texas State University, San Marcos, TX.

Diamond, D.D., Riskind, D.H. 1987. The natural vegetation of Texas: A preliminary classification. *TX. J. Sci.* in press.

Diamond, J.M. 1975. The island dilemma: Lessons of modern biogeographic studies for the design of nature reserves. *Biol. Conserv.* 8:73–88.

Echelle, A.A. 1975. A multivariate analysis of variation in an endangered fish, *Cyprinodon elegans*, with an assessment of populational status. *TX. J. Sci.* 26:529–538.

Everitt, J.H., Gonzales, C.L. 1987. Botanical composition of eleven south Texas range sites. *In* D.H. Riskind, and G.W. Blacklock (eds.), *Symposium on the Tamaulipan Biotic Province.* Texas Parks and Wildlife Department, Austin, TX. In press.

Fonteyn, P.J., Stone, M.W., Yancy, M.A., Baccus, J.T. 1984. Interspecific and in-
traspecific microhabitat temperature variations during a fire. *Am. Midland Naturalist*
112:246–250.

Frentress, C.D., Frye, R.G. 1975. Wildlife management by habitat units—a preliminary
plan of action, pp. 245–262. *In First Earth Resource Survey Symposium Proceedings.*
National Aeronautics and Space Administration, Houston, TX.

Fulbright, T. 1985. Germination of anacua, spiny hackberry, and brasil seeds. Final
report prepared for Texas Parks and Wildlife Department by Texas A&I University,
Kingsville, TX.

George, R.R. 1985. White-winged dove management in Texas with implications for
northwestern Mexico. Presented at First Regional Conference of the Rio Grande
Border States on Parks and Wildlife, Laredo, TX.

Gould, F.W., Hoffman, G.O., Rechenthin, C.A. 1960. Vegetational Areas of Texas.
Texas Agricultural Experiment Station Leaflet no. 492, Texas A&M University, College Station, TX.

Graber, J.W. 1961. Distribution, habitat requirements, and life history of the black-
capped vireo *(Vireo atricapillus). Ecol. Monogr.* 31:313–336.

Haas, R.H., McGuire, J.M. 1974. Evaluating remote sensing applications for land resource management: Phase II summary report. Remote Sensing Center, Texas A&M
University, College Station, TX.

Harris, L.D. 1984. *The Fragmented Forest.* University of Chicago Press, Chicago, IL.

Hendrickson, D.A., Minkley, W.L. 1985. Ciénegas—vanishing climax communities of
the American Southwest. *Desert Plants* 6(3):1–175.

Higgs, A.J., Usher, M.B. 1980. Should nature reserves be large or small? *Nature* 285:568–
569.

Hubbs, C., Echelle, A.A. 1972. Endangered non-game fishes of the Upper Rio Grande
Basin, pp. 147–167. *In Symposium on Rare and Endangered Wildlife of the Southwestern United States.* New Mexico Department of Game and Fish, Santa Fe, NM.

Janzen, D.H. 1983. No park is an island: Increase in interference from outside as park
size decreases. *Oikos* 41:402–410.

Johnston, M.C. 1963. Past and present grasslands of southern Texas and northeastern
Mexico. *Ecology* 44:456–466.

Kier R.S., Garner, L.E., Brown, Jr., L.F. 1977. Land resources of Texas—a map of
Texas lands classified according to natural suitability and use considerations. Bureau
of Economic Geology, University of Texas, Austin, TX.

Kingston, M. (ed.) 1985. *1986–1987 Texas Almanac.* The Dallas Morning News, Dallas,
TX.

Kroll, J.C. 1974. Nesting success of the golden-cheeked warbler *(Dendroica chrysoparia)*
in relation to manipulation of Ashe juniper *(Juniperus ashei)* habitat. Progress report,
U.S. Fish and Wildlife Service, Bureau of Sport Fisheries and Wildlife, Albuquerque,
NM.

Kroll, J.C. 1980. Habitat requirements of the golden-cheeked warbler: Management
implications. *J. Range Mgmt.* 33:60–65.

Leopold, A.S., Cain, S.A., Cottam, C.H., Gabrielson, I.N., Kimball, T.L. 1963. Wildlife
management in the national parks. *Am. Forests* 69:32–35, 61–63.

Margules, C., Higgs, A.J., Rafe, R.W. 1982. Modern biogeographic theory: Are there
any lessons for nature reserve design? *Biol. Conserv.* 24:115–128.

Marshall, J.T., Clapp, R.B., Grzybowski, J.A. 1985. Status report: *Vireo atricapillus.*
Office of Endangered Species, U.S. Fish and Wildlife Service, Albuquerque, NM.

May, R.M. 1975. Island biogeography and the design of wildlife preserves. *Nature*
254:177–178.

McMahan, C.A., Frye, R.G., Brown, K.L. 1984. The vegetation types of Texas, including
cropland-an illustrated synopsis and map. Federal Aid Project W-107-R, Texas Parks
and Wildlife Department, Austin, TX.

Miller, R.R. 1961. Man and the changing fish fauna of the American Southwest. Papers of the Michigan Academy of Science, Arts, and Letters 46:365–404.

Neck, R.W., Everitt, J.H., Gerbermann, A.H. 1984. Native brush communities associated with resacas in the Lower Rio Grande Valley of Texas. Paper presented at annual meeting, Texas Academy of Science, San Antonio, TX.

Pace, L., Garrison, J.C., Davis, A., Riskind, D. 1975. Helping nature recover. *TX. Parks Wildlife Mag.* 33:6–11.

Peters, R.L., Darling, J.D.S. 1985. The greenhouse effect and nature preserves. *BioSci.* 35:707–717.

Pickett, S.T.A., Thompson, J.N. 1978. Patch dynamics and the design of nature reserves. *Biol. Conserv.* 13:27–37.

Polunin, N., Eidsvik, H.K. 1979. Ecological principles of the establishment and management of natural parks and preserves. *Environ. Conserv.* 6:21–26.

Pulich, W.M. 1962. In quest of the golden-cheeked warbler. *TX. Ornithol. Soc. Newslet.* 10:5–11.

Pulich, W.M. 1976. The golden-cheeked warbler: A bioecological study. Texas Parks and Wildlife Department, Austin, TX.

Riskind, D.H. 1977. Natural resource management in Texas parklands: An overview. *Bull. TX. Ornithol. Soc.* 10:26–30.

Simberloff, D.S., Abele, L.G. 1976. Island biogeographic theory and conservation practice. *Science* 191:285–286.

Soulé, M.E. 1985. What is conservation biology? *BioSci.* 35(11):727–734.

Soulé, M.E., Simberloff, D. 1986. What do genetics and ecology tell us about the design of nature reserves? *Biol. Conserv.* 35:19–40.

Stankey, G.H., Cole, D.N., Lucas, R.C., Petersen, M.E., Frissell, S.S. 1985. The limits of acceptable change (LAC) system for wilderness planning. General Technical Report INT-176, USDA Forest Service, Intermountain Forest and Range Experiment Station, Ogden, UT.

Stone, M.W. 1984. Microhabitat temperature variation and the post-burn response of vegetation in the Edwards Plateau of Texas. Master's thesis, Southwest Texas State University, San Marcos, TX.

Tewes, M.E., Everett, D.D. 1982. Status and distribution of ocelot and jaguarundi in Texas. *In* International Cat Symposium Proceedings, Cesar Kleberg Wildlife Research Institute, Texas A&I University, Kingsville, TX.

Texas Parks and Wildlife Department. 1976. Job Performance Report. Job no. 1, Federal Aid Project W-107-R, Segment 2, Texas Parks and Wildlife Department, Austin, TX.

Texas Parks and Wildlife Department. 1978. Job Performance Report. Job no. 1, Federal Aid Project W-107-R, Segment 4, Texas Parks and Wildlife Department, Austin, TX.

Texas Parks and Wildlife Department. 1980. Job Performance Report. Job no. 1, Federal Aid Project W-107-R, Segment 6, Texas Parks and Wildlife Department, Austin, TX.

Texas Parks and Wildlife Department. 1981. Job Performance Report. Job no. 1, Federal Aid Project W-107-R, Segment 7, Texas Parks and Wildlife Department, Austin, TX.

United States Fish and Wildlife Service. 1985. Land protection plan for Lower Rio Grande Valley National Wildlife Refuge in Cameron, Hidalgo, Starr, and Willacy counties, Texas. Southwest Region, Fish and Wildlife Service, United States Department of the Interior, Albuquerque, NM.

Waggerman, G.L. 1978. Experimental restoration of white-winged dove nesting habitat. Pittman-Robertson Project W-78-D, Texas Parks and Wildlife Department, Austin, TX.

Whitcomb, R.F., Lynch, J.F., Obler, P.A., Robbins, C.S. 1976. Island biogeography and conservation: Strategy and limitations. *Science* 193:1030–1032.

White, P. 1986. Many *and* large, large *and* small: Nature reserves debate goes on. *Park Sci.* 6:13–15.

Williams, M.H. 1975. The design of wildlife reserves. *Nature* 256:519.

11. Progressiveness Among Farmers as a Factor in Heterogeneity of Farmed Landscapes

Joan Iverson Nassauer and Richard Westmacott

11.1 Farming and Landscape Heterogeneity

Human attitudes and actions, broadly interpreted, are fundamental factors in landscape ecology and the management of disturbance. The obvious reason is that humans have always had, and continue to have, a major impact on landscapes. Humans may be the source of disturbance (such as in the generation of air pollution; see Bormann, Chapter 3), but they may also create landscapes (see Chapters 7, 8, 9, and 10). Human action on the land is the result of attitudes derived from a complex of ideas, motivations, and experiences. If, as Risser (Chapter 1) suggests, humans are to be included within landscape ecology, then we must include within our studies the attitudes and motivations underlying human action in creating or responding to landscape disturbance. The role of farming in the rural agricultural landscape provides an example of this phenomenon.

Farming allows ideals and values to be directly translated into land management decisions. The farmer identifies with the agricultural landscape, and this landscape represents the farmer. A farmer's work is constantly on view, and the farmer's care of the land can be readily judged by his peers. Consequently, the agricultural landscape becomes a display of the farmer's knowledge, values, and work ethic.

Concern has been voiced that changes in farming practices have decreased heterogeneity in rural landscapes (Westmacott and Worthington 1974; Todd

1976) or caused, as J.B. Jackson describes it, "a coarseness of detail" (Zube and Zube 1977). The loss is not only of biological diversity but also of economic, social, and visual diversity. Carlson (1985) expresses this idea well:

> Not only have the fields become vast flat tracts of land exclusively devoted to a single crop, they have become devoid of many traditional features of the rural landscape. In the quest for large uniform farming surfaces, topographical irregularities such as gullies, washes, sloughs, rises, slopes and knolls have succumbed to land remodelling. At the same time features once essential to rural life such as woodlots, windbreaks, ponds, fences, country schools, rural churches, and outlying farm buildings are systematically being removed or destroyed.

In 1942, both farmers and the general public in Britain believed that good and prosperous farming would result in "good" landscape (Scott 1942). Mixed rotational farming had created a landscape which was interesting, visually satisfying, and biologically diverse. In the United States, Bromfield (1950) wrote of traditional farm landscapes which had become part of America's heritage, and he, too, attributed "good" landscape to "good" farming:

> The farmer may leave his stamp on the whole of the landscape seen from his window, and it can be as great and beautiful a creation as Michelangelo's David, for the farmer who takes over a desolate farm, ruined by some evil and ignorant predecessor, and turns it into a paradise of beauty and abundance is one of the greatest of artists.

At the turn of the century, the techniques associated with "good" farming were quite unlike those used today. In more recent years, government, agribusiness, and the land grant university systems of the United States have advocated new farming techniques. These innovations were recommended primarily to increase crop yields, and the holistic notion of "good" landscape was not a primary objective. The subsequent loss of farmed landscape heterogeneity is widely apparent.

The objective of this chapter is to examine some of the personal values of farmers and show how aspects of their value system can influence the landscape. We review three separate but similar studies. Farmers' attitudes toward the rural landscape were determined in Illinois (midwestern United States), lowland areas of England, and upland areas of England and Wales by interviewing farm operators and (in Illinois) asking them to judge pictures of rural landscapes. These interviews suggested that one factor in the rural value system was the concept of progressiveness. Within the context of landscape as a display of self, the farmer wishes to be seen as progressive. To maintain such an image, farmers frequently take actions which cannot be justified economically. Typically, a farmer might admire an agricultural landscape with which he was not familiar and describe the farmer who worked the land as progressive. Or, he might explain his decision to alter a historical feature of his own farm as consistent with his progressiveness.

In most interviews, progressiveness represented values that imply a positive disposition toward change that supports better farming. Defined in this way, progressiveness may be strongly related to frequent disturbance of the rural landscape, altering the pattern of agricultural and nonagricultural elements. However, progressiveness may be positively or negatively related to heterogeneity depending on the progressive technique adopted. Farmers in our Illinois survey, for example, identified minimum tillage techniques with progressiveness. The particular farming techniques that farmers believe to be progressive, rather than progressiveness alone, may directly affect heterogeneity. Progressiveness is a dynamic concept, and the elements of farming practice that are considered progressive change through time and across cultures and, therefore, this concept may positively or negatively affect landscape properties.

11.2   Progressiveness and Innovation

We define progressiveness, a word used by both British and American farmers who were interviewed, as a positive disposition toward change that is believed to support "good" farming. This is generally consistent with the *Oxford English Dictionary* (OED) definition "growing, increasing, developing; usually in a good sense; advancing towards better conditions" (OED 1933). It may not be co-incidental tht one of the most influential farming periodicals in the United States is *The Progressive Farmer*.

Clearly at issue is the farmers' definition of good farming. We suggest that at the turn of the century the definition accommodated and may even have included a qualitative and holistic sense of landscape; however, in more recent times, the persuasion of farmers has been toward a narrower and more short-term view.

The adoption of modern farming practices by farmers (termed the diffusion of innovation) has been a major topic of study for rural sociologists. In this literature, concepts similar to progressiveness have been found to predict the diffusion of innovation. For example, Pampel and Van Es (1977) found three factors that predicted the adoption of potentially profitable commercial practices: a psychological factor including receptivity to new ideas, a belief in farming as a way of life, and the promise of higher profitability.

At the same time, the literature underscores the importance of economic factors, including farm size, in the adoption of innovation. Buttel and Larson (1979) found a consistent positive association between farm size and energy intensity of crop production. They also found that small-scale farmers and those with a noneconomic orientation to farming were generally supportive of soil conservation (Buttel et al. 1981). However, they note that a supportive attitude does not necessarily result in the adoption of innovative practices. It has been suggested recently that although small farmers may have attitudes as progressive as larger farmers, they may not actually adopt innovative practices because of various institutional or environmental constraints (Hefferman and Green 1986). Progressiveness should be seen as one of a constellation of factors related to the adoption of agricultural innovation.

11.3 Progressiveness and Aesthetic Response

Farmers' perception of "good farming" is not a purely economic assessment, but is in part an aesthetic response. Farmers see a beauty in rural landscapes that is rooted in their understanding of the land's function, the fit between its economically productive use and its suitability. Landscapes suitable to be cropland, by virtue of their soil, slope, and locational characteristics, are beautiful when they fulfill their purpose well. In the Midwest, this is exemplified when rows are straight and even, when the field is large and flat and uninterrupted, and when good resource stewardship is apparently being practiced. The ugly landscape is one that is used in a way that contradicts its suitability. A crop being raised on wet, weedy bottomland is not beautiful despite its picturesque riverine setting. Similarly, eroding cultivated slopes are not beautiful, even in the foreground of panoramic views.

For viewers who do not understand agricultural production or landscape suitability as farmers typically do, these same scenes might be quite pleasant. In 1889, William James described a North Carolina farm as a scene of "unmitigated squalor." Yet, he recognized that when farmers looked at the rough clearings with stumps and girdled trees "which to me was a mere ugly picture . . . (it) was to them a symbol redolent with moral memories and sang a very paean of duty, struggle, and success" (Carlson 1985).

Thus, aesthetic response is not a whimsical, mysteriously personal experience, but a powerful product of environmental context and viewer understanding. The rural landscape is viewed differently by farmers and urban people, partly because of different levels of knowledge about the environment. Farmers enjoy seeing "good" farming, not only because it represents an economic return but because it has an aesthetic value as well. Reflecting on the fact that 9 of 10 people were working in agriculture when the United States was founded, John Stilgoe suggests "that much of our national heritage subtly emphasizes the good life of husbandry and the beauty and rightness of land shaped for farming" (Stilgoe 1982).

Socrates argued that all things were good and beautiful in relation to those purposes for which they are well adapted (Hipple 1957). In 18th Century Britain, David Hume expanded on the concept of "utilitarian beauty" as applied to farmed landscapes.[1] "Nothing renders a field more agreeable than its fertility, . . . " he wrote, "I know not but a plain, overgrown with furze or broom may be in itself as beautiful as a hill cover'd with vines or olive trees, tho' it would never appear so to one who is acquainted with the value of each" (Hipple 1957).

Farmers respond to rural landscapes with a very similar utilitarian sense of beauty. Just as utility changes with new technology and new information, farm-

---

[1] In the 18th Century, utilitarian beauty was not quite the same as functional beauty. Edmund Burke for instance discusses "fitness" and, although he derives pleasure from contemplating an object perfectly fitted for its task, he denies that it can be called beauty. "The snout of the hog is not lovely because it is adapted to its office" (Hipple 1957).

ers' aesthetic response to farmed landscapes is likely to change as new concepts of functional agriculture are accepted. Progressiveness, as revealed in farmers' land, becomes both cause and effect of this dynamic aesthetic response. Farmers like to see landscapes that look progressive, and, at the same time, progressive landscapes suggest new images of landscape beauty.

11.4 Progressiveness Among Farmers

11.4.1 Hypotheses

Our hypothesis is that the appearance of progressiveness, as revealed in the farmed landscape, is important to most farmers. Further, we hypothesize that actions which are thought to demonstrate progressiveness are taken, even in the full knowledge that they may not be profitable.

11.4.2 Illinois Study

Interviews were held with 17 Illinois farmers, selected to represent different types of operations (dairy, beef, hog, and grain). Some farmed directly adjacent to expanding urban areas, and some were well removed from urban expansion. Slides of 10 different countryside landscapes were shown to the farmers in their homes. These were color slides taken with a 50-mm lens. Each landscape was shown at four or five different times during the year to represent seasonal change. The farmers were asked open-ended questions about what they noticed and found attractive and unattractive about each view. Interviews typically lasted three hours.

The farmers interviewed consistently identified apparently progressive management with attractive landscapes. . Their comments suggest that when looking at agricultural landscapes, farmers want to see change. Their comments suggest that concepts that also influence farmers' perception of landscape attractiveness are: productivity, orderliness, and care. Specific landscape elements that were associated with these concepts were: mown lawns and roadsides, pruned trees, weed-free fields, straight rows of crops, new or newly painted buildings, and, interestingly, minimum tillage techniques.

Farmers have always "improved the land," adding to its value by clearing it of trees and stones and by building structures. The progressive farmer has traditionally fought back unruly nature to make his land productive. Responding to a slide of a nature study area (Fig. 11.1), one farmer made this typical remark: "I would like to get the pruning shears out. This is too much of a good thing. It has an old-fashioned look."

Farmers claim the land for orderly production. Commenting on the trees growing along a channelized stream (Fig. 11.2), another person said, "A farmer fights back that type of thing. That would be a forest but would never yield anything of value." Looking at a different landscape, another farmer said, "This is one of those places where the woods will take 5 feet from you every year if you let it."

Figure 11.1. Nature study area.

Figure 11.2. Channelized stream.

Along with this idea of improvement comes an appreciation for neatness. Neatness begins to be valued outside of its direct economic implications. Unpainted buildings, old rusting machinery, unmown roadsides, unpruned ornamental trees, or weedy fallow fields may not reduce productivity; however, they are likely to give the impression that a farmer does not care and is not improving the land. Looking at a farm which she saw as very neat, one farmer commented on its opposite, "If the buildings are falling down, the roadside isn't taken care of."

More recently, farmers have shown their progressiveness by using "advanced" scientific technologies ranging from applied economics to chemical fertilizers. Progressiveness has come to be identified not only with ordering nature and creating a neat, clean landscape, but also with a willingness to innovate. Farmers' responses indicated that they equated newness and uniformity with functionalism (Fig. 11.3). "That old barn stands out. It looks out of place next to the other farm buildings. It's just not functional anymore." "This is very functional, all but one building has been replaced." "The barn doesn't fit with the rest. It should be painted, sided, or B-and-B'ed (burned and bulldozed)." The farmers were impressed with the neatness of the same farm, in general. "This is well-maintained. Probably owner operators." "The buildings are quite new and modern—useful." "No fences. No livestock. This fellow must be progressive."

The definition of the progressive ideal must be tempered with the expectation that farmers' notions of improvement and advancement will change with their knowledge. For instance, several farmers commented that they did not like the unkempt look of dead trees, which were present in the nature study area (Fig. 11.1). However, one said, "That's pretty, but there are some dead trees. I *used to* dislike dead trees. It depends on if its part of a natural process." As ecological processes become better understood, advanced farming techniques are likely to change. The interviewees often expressed admiration for good stewardship, and their concepts of stewardship appeared to vary with their knowledge. Seeing minimum tillage for soil conservation in a fall-plowed field (Fig. 11.4), one said, "Knowing that's chisel-plowed makes that look better."

11.4.3   England and Wales Studies

A study of farmers in lowland agricultural areas of England carried out in 1972 (Westmacott and Worthington 1974) showed their values to be remarkably similar to the Illinois farmers. The approach used in this study was to determine what physical changes had been made on farms in the last two or three decades and then to discuss with the farmers the reasons these changes had been made. In cases where farmers had not made changes which most of their neighbors had made, they were asked why not.

Field surveys, aerial photographs, and interviews were used. Eight or more contiguous farms in each of seven lowland areas, which were broadly representative of different farming systems and physiography, were surveyed. The

Figure 11.3. Older barn with new buildings.

Figure 11.4. Minimum tillage in fall-plowed field.

responses were therefore those of quite a small sample of farmers but were confirmed by other farmers and farm organizations during the course of the study.

The interview questionnaire did not contain any items specifically aimed to assess progressiveness. However, farmers used this term to describe why changes had been made on the farm. Differences between farmers' evaluations of "good" landscape and those of naturalists also became clear. Hedgerows, hedgerow trees, ponds, open drainage ditches, and other, usually defunct, components of the hetrogeneous, traditional mixed farm landscape were frequently removed even when the farmer could not justify the action on economic grounds. For instance, hedgerows were sometimes removed not to facilitate mechanization or to effect changes in grazing management but because they were derelict and viewed as a sign of backwardness. Several Fenland farmers expressed the opinion that trees were "untidy clutter" and spoiled the appearance of their crops. In other areas most saw hedgerow trees as causing nuisance or inconvenience. In fact, on large farms growing a single crop, such as wheat, hedgerows become obstructions to new large-scale machinery. In Huntingdonshire only 36.5% of the land was tillage crops in 1936 compared with 70% today, mostly in wheat. Between 1945 and 1983, 37 m per hectare of hedgerow has been removed; on a 500-acre farm, that is 3.7 miles of hedgerow.

Farmers were asked whether they thought that farming had been "responsible for the beauty of the countryside and whether society should accept whatever modern agriculture produces." Responses to this question emphasized that the belief that "good" farming produces "good" landscape was still alive in 1972. To most farmers however, "good" farming is progressive farming and in different farming systems, there are different signs of progressiveness. In the highly specialized cereal growing areas in the east, where economies of scale have the greatest potential financial benefit, wide open landscapes, clean and uniform crops, and stark and pristine farm buidings are signs of a progressive farm. Tramlines, tracks left in the growing crop by regularly using the same route for the boom sprayer, give the fields a mechanical precision in which the farmer takes great pride. Farmers in these areas are least tolerant of trees and defunct hedges. In contrast, in areas where mixed farming is still practiced by the majority of farmers, many of the features of mixed rotational farming are still functional and therefore their continued existence is not a sign of backwardness. In specialized livestock areas, the most important sign of progressiveness is the greenness of the grass. Short-term leys, usually rye-grass with heavy additions of lime and nitrogenous fertilizer, replace herb-rich meadows. The most progressive farmers will replace hedgerows with wire, dividing the farm into small paddocks.

A similar examination of changing landscapes in upland areas of England and Wales (Westmacott et al. 1977; Sinclair 1983) showed that farmers in hill country have quite different attitudes from their lowland counterparts. The method was similar to that used in the lowland research but a larger sample was used, 10 to 15 farmers in each of 12 different areas. The data suggest that changes in farm management are perceived as a significant causal factor in landscape change. However, only 27% of the farmers interviewed thought that the changes *on their own farms* had produced a significant impact on the landscape. Of these, 89% thought that the changes they had made had "improved" the landscape. It is interesting however to note that these improvements were generally described as "tidying up" or "improving" the land, thereby improving the scenery. A large number of farmers pointed out how much greener their hills were now than previously. A farmer in Glascwm (central Wales), where moorland had been converted to grassland said proudly "reclamation has turned the scene green where there was nothing but grey" (Westmacott et al. 1977). The transformation gives rise to quite different sentiments from the naturalist.

Although few farmers saw their changes as having significant impact on the landscape, "an overwhelming 97% recognized that farming plays an important part in the appearance of the landscape" and more than one-quarter of those interviewed went further and stated that "farming was the countryside and vice versa" (Sinclair 1983). Clearly, tidiness and greenness were two visible characteristics of the farm that upland farmers believed were important to their "good farmer" image and, mistakenly, to their "good steward" image. But, in the uplands far fewer farmers pointed to examples of agricultural progressiveness as improving the landscape. Indeed, when asked what they found attractive about their area, more farmers cited the *variety* of scenery than any other characteristic. This variety is clearly more typical of rugged upland areas than of more intensively farmed lowland areas. Although lowland farmers were not asked the same question, uniformity of crop was frequently mentioned as enhancing the landscape. Agriculturally unproductive areas, , which add variety to the landscape, were seen as functional obstructions to farming and visual detractors.

In the upland areas, which have seen some of the most bitter confrontations between farming and conservation interests, a majority (69%) of farmers interviewed recognized that farmers "had a responsibility for more than food production." This was seen primarily in terms of stewardship—the idea the farmers had a "responsibility to look after the countryside . . . " (Sinclair, 1983). When asked whether they thought farmers in general should be prepared to modify their methods in the interests of conserving the landscape and wildlife, a majority (57%) still gave an unqualified yes. Even the most pessimistic interpretation of this and other questions suggests that nearly half the upland farmers were prepared to forego some profit for the sake of landscape conservation. In contrast, between six and eight years earlier, a majority of lowland farmers had stated that they would be unwilling to modify their farming practices unless compensation was to be paid (Westmacott et al. 1977). However, younger upland farmers, especially those with large holdings were more likely than older farmers to be unwilling to make any concessions to conservation in their farming practices.

11.5 Progressiveness and Landscape Heterogeneity

These studies suggest that progressiveness is a widely held value among farmers and that progressiveness affects farmland management decisions. However, progressiveness should not be understood as a disposition toward a particular technique or approach but rather a disposition for change that supports "good" farming. Farmers in cultures that are similar but not identical shared this ideal. To be a progressive farmer is to be a good farmer. To be a progressive farmer is to be ready to change.

Information about innovation as well as understanding of traditional techniques are essential before the farmer adopts any particular technique. If farmers have sufficient information about the origin and management of hedgerows (Forman and Baudry 1984), for instance, their progressiveness may motivate them to create and maintain hedgerows in the next 50 years just as it motivated them to remove hedgerows in the past 50.

Progressiveness is directly related to landscape disturbance; progressive farmers are likely to introduce new land management regimes. However, progressiveness is only indirectly related to landscape heterogeneity. The nature of the innovations adopted by progressive farmers in pursuit of "good" farming will determine whether heterogeneity decreases or increases with new land management regimes. Progressiveness is one of a number of conditions, including economic factors, which might motivate a farmer to adopt innovations. Future research must recognize the distinction between progressiveness as an ideal and actual progressive behavior.

Postwar changes in farmland management have progressed along a very limited dimension. Landscape change that follows from a more holistic concept of landscape productivity can also be framed as progressive. The progressive ideal changes with new information. Translating landscape ecology into everyday farming practices offers great potential for change.

References

Bromfield, L. 1950. *Out of the Earth*. Harper, New York.

Buttel, F.H., Larson, III., O.W. 1979. Farm size, structure, and energy intensity: An ecological analysis of U.S. agriculture. *Rural Sociol.* 44:471–488.

Buttel, F.H., Willespie, G.W., Larson, III, O.W., Harris, C.K. 1981. The social bases of agrarian environmentalism: A comparative analysis of New York and Michigan farm operators. *Rural Sociol.* 46:391–410.

Carlson, A. 1985. On appreciating agricultural landscapes. *J. aesthetics art crit.* 43:301–312.

Forman, R.T.T., Baudry, J. 1984. Hedgerows and hedgerow networks in landscape ecology. *Environ. Mgmt.* 8:495–510.

Hefferman, W.D., Green, G.P. 1986. Farm size and soil loss: Prospects for a sustainable agriculture. *Rural Sociol.* 51:31–42.

Hipple, J.W. 1957. *The Beautiful, the Sublime, and the Picturesque in Eighteenth Century British Aesthetic Theory*. Southern Illinois University Press, Carbondale, IL.

Oxford English Dictionary 8. 1933. Clarendon Press, Oxford, United Kingdom.

Pampel, Jr., F., Van Es, J.C. 1977. Environmental quality and issues of adoption re-
    search. *Rural Sociol.* 42:57–71.
Scott, J. 1942. Report of the committee on land utilization in rural areas. Cmnd. 6378.
    Her Majesty's Printing Office, London.
Sinclair, G. 1983. The upland landscapes study. Environmental Information Services,
    Martlewy, Dyfed, Wales.
Stilgoe, J.R. 1982. *Common Landscapes of America, 1580–1845.* Yale University Press,
    New Haven, CT.
Todd, J. 1976. A modest proposal: Science for the people. *In* R. Merrill (ed.), *Radical
    agriculture.* Harper, New York.
Westmacott, R., Worthington, T. 1974. New agricultural landscapes. The Countryside
    Commission for England and Wales, Cheltenham.
Westmacott, R., Bell, S., Sinclair, G. 1977. The upland landscape study: Pilot study
    areas. Unpublished report to the Countryside Commission for England and Wales.
Zube, E., Zube, M.J. 1977. *Changing Rural Landscapes.* University of Massachusetts
    Press, Amherst, MA.

4. Conclusion

12. The Ethics of Isolation, the Spread of Disturbance, and Landscape Ecology

Richard T.T. Forman

Three relatively distinct major themes emerge from the chapters of this book. I will first introduce and briefly explore a key question of ethics, as this is a foundation of all planning, management, and individual actions in landscapes. The second section will focus on how landscape heterogeneity affects the spread of disturbance, the subject of this book and a 1986 symposium held in Athens, Georgia. The final section considers the growth of landscape ecology, as both the conceptual and applied framework for this volume.

12.1 The Ethics of Isolation

To simplify we isolate. Few of us can visualize the effects of three factors varying together; indeed only a rare human being can envision four covarying factors. Yet major issues and problems hinge on many variables, often with several being of major significance.

So our approach is to eliminate and ignore most variables as minor, and propose a solution to an issue based on one or a few of the most important variables. Thus, despite knowing better, the politician can answer "yes" or "no" to a complex question, and the decision-maker can argue a problem will no longer exist if we ameliorate the one or two leading variables. In this way our collective conscience is salved. But is this ethical?

The challenge crystallizes when considering human actions on land. Should an industry knowingly "externalize" its costs on society by lowering the surrounding water table for its cooling processes? Should a planner design a housing project that causes roads of an adjoining town to be rebuilt due to commuter traffic jams? Should a wildlife manager maintain high herbivore populations that overuse adjacent farmers' land? Should a farmer use heavy pesticide applications that blow or wash into an adjacent wildlife refuge or natural area? In each case a certain piece of land is considered in isolation, independent of its surroundings. These are examples of bad planning, but are they unethical?

Let us examine this overarching and critical issue using the lessons of "biophilia" (Wilson 1984), "Plato's land alarm" (Critias 111), and the "land ethic" (Leopold 1949). We then add several concepts from landscape ecology, and see the ethics of isolation come together in a general principle.

12.1.1 Biophilia and Plato's Land Alarm

An *ethic* is a self-imposed limitation on the freedom to act. Ethics constrain many human endeavors, including economic (e.g., cooperation in lieu of competition), social (e.g., the rarity of mass murder), and ecological (e.g., protection of eagle habitat and panda bears). Although I am not an ethicist, I would like to briefly explore some relevant spatial and temporal issues in an ecological context, in the hope that applied landscape ecology will lead to a substantial and noticeable improvement in our landscapes.

Wilson (1984) defines *biophilia* as "the innate tendency to focus on life and life like processes." He points out that " . . . to explore and affiliate with life is a deep and complicated process in mental development. To an extent still undervalued in philosophy and religion, our existence depends on this propensity, our spirit is woven from it, hope rises on its currents." Put another way, he argues that:

> . . . we are human in good part because of the particular way we affiliate with other organisms. They are the matrix in which the human mind originated and is permanently rooted, and they offer the challenge and freedom innately sought. To the extent that each person can feel like a naturalist, the old excitement of the untrampled world will be regained. I offer this as a formula of reenchantment to invigorate poetry and myth: mysterious and little known organisms live within walking distance of where you sit. Splendor awaits in minute proportions.

This dynamic view of the inextricable linkage humans have with other life is rooted in evolution, includes ecology and human physiology, and leads to the human spirit. Here we cannot avoid ethics. If we incrementally destroy other life, we not only progressively weaken our life support system, but simultaneously amputate both our roots and our spirit (Fig. 12.1). Recognition of these connections of humans with other life leads to a powerful ethic.

As a product of human evolution we are a walking configuration of nerve cells. A genetic component underlies not only the nerve system structure, but also our reasoning, including moral reasoning (Ruse and Wilson 1985; E.O. Wilson, personal communication). We cannot consciously alter this ethical

Figure 12.1. Mount Kilimanjaro in Tanzania, Africa. Humans who evolved in savannas and forests "in its shadow" still depend on its plants, animals, soil, and water, and are spiritually nourished by its presence. The linkage between humans and other life, as embodied in biophilia, has both temporal and spatial dimensions. (Photo courtesy of W. C. Johnson.)

foundation, and are constrained by it. Yet ethics change. An individual or a society gradually alters its ethics on many matters, including economics, health, shelter, food, and landscape quality. The environment and learning provide a mechanism for these changes in a self-imposed limitation on the freedom to act.

Let us examine an early account of a society-land interaction. Plato (ca. 360 BC, Critias, 111.b, c, d) describes a process of human history repeated well into the present:

> . . . (I)n those days the country . . . yielded far more abundant produce. . . . (I)n comparison of what then was, there are remaining only the bones of the wasted body, as they may be called. . . . all the richer and softer parts of the soil having fallen away, and the mere skeleton of the land being left. . . . Moreover, the land reaped the benefit of the annual rainfall, not as now losing the water which flows off the bare earth into the sea, but, having an abundant supply in all places, and receiving it into herself and treasuring it up in the close clay soil, it let off into the hollows the streams which it absorbed from the heights, providing everywhere

abundant fountains and rivers, of which there may still be observed sacred memorials in places where fountains once existed. . . .

According to this account, by the 4th Century BC humans had already ravaged the Greek land. *Plato's land alarm,* a warning cry to the people that overuse of the land leads to land degradation, is a timeless message. This process of landscape change, doubtless largely caused by vegetation loss and excessive animal grazing, brings into focus the need for an affiliation and respect of humans for living organisms, the essence of biophilia.

12.1.2   The Land Ethic

What solutions can prevent the all too frequent land degradation process? Leopold (1949) addressed this difficult question and suggested three possibilities. One is a legal solution. Governments pass laws protecting game, parks, or natural areas. Such laws work if coupled with enforcement, and if not too far out of step with the prevailing social ethics of what people think they can do to the land. However, this is like a thumb in the dike. "It tends to relegate to government many functions eventually too large, too complex, or too widely dispersed to be performed by government" (Leopold 1949).

A second possible solution is economic self-interest. A timber company protects the land to harvest wood products, or a landowner protects the land to have a private hunting and fishing reserve. Again Leopold pinpoints the flaw: " . . . a system of conservation based solely on economic self-interest is hopelessly lopsided. It tends to ignore, and thus eventually to eliminate many elements in the land community that lack commercial value, but that are (as far as we know) essential to its healthy functioning. It assumes, falsely, I think, that the economic parts of the biotic clock will function without the uneconomic parts."

A third possible solution to landscape protection is called the *land ethic.* Leopold describes it in the following ways.

> An ethical obligation on the part of the private owner is the only visible remedy . . .

> We abuse land because we regard it as a commodity belonging to us.

> That land is a community is the basic concept of ecology, but that land is to be loved and respected is an extension of ethics. That land yields a cultural harvest is a fact long known, but latterly often forgotten.

"An ethic, ecologically, is a limitation on freedom of action in the struggle for existence. An ethic, philosophically, is a differentiation of social from antisocial behavior. These are two definitions of one thing. The thing has its origin in the tendency of interdependent individuals or groups to evolve modes of cooperation. The ecologist calls these symbioses." Note the parallel here with biophilia. All life, and the only life we know, has evolved together here on planet Earth. Modes of cooperation or complementarity are manifold. That humanity could live on a biologically impoverished land is but an illusion.

Leopold continues:

> The land ethic simply enlarges the boundaries of the community to include soils, waters, plants, and animals, or collectively: the land.

> This sounds simple: do we not already sing our love for and obligation to the land of the free and the home of the brave? Yes, but just what and whom do we love? Certainly not the soil, which we are sending helter-skelter downriver. Certainly not the waters, which we assume have no function except to turn turbines, float barges, and carry off sewage. Certainly not the plants, of which we exterminate whole communities without batting an eye. Certainly not the animals, of which we have extirpated many of the largest and most beautiful species.

Thus, we see life and the land as one. We evolved in this interlocking system and are dependent on it for our livelihood; we are nourished by it for our spirit (Fig. 12.1). Government and economic self-interest are not up to the task of adequate landscape protection. Eloquently described nearly four decades ago, a satisfactory future for the land and for us depends primarily on a land ethic.

Why then, we must ask, does the land continue to deteriorate? Two elements appear critical. First, when separated from economics, a land ethic is too idealistic, too much the domain of the rich. It works in scattered local areas, but rarely in whole landscapes. Second, the bulk of the earth's population is concerned primarily with immediate and adequate food, shelter, and health. Long-term land protection is of little concern and is usually considered inimical to solving the immediate critical needs.

In view of these two major constraints, I believe the key to arresting landscape degradation may be an integration of the three apparently disparate elements, governmental regulation, economic self-interest, and the land ethic. As noted, ethics change as society continually grapples with right and wrong. Thus, if governmental regulation began to always include a consideration of the long-term effect of an action on the land, a feedback effect on the private individual and the corporate body to do the same would certainly evolve. The results of governmental and economic actions would then gradually approach the tenets and benefits of the land ethic.

Clearly a catalyst is needed to convert idealistic approaches to arresting landscape deterioration into realistic approaches. Such catalysis, however, must have a spatial framework. Indeed, a few key principles of landscape ecology appear to provide this framework.

12.1.3   Isolation and Landscape Ecology

We have learned that a landscape may be usefully defined scientifically as a heterogeneous land area composed of a cluster of interacting ecosystems that is repeated in similar form throughout. Landscapes vary in size down to a few square kilometers. Central to this concept is the existence of a *cluster of ecosystems* found throughout the landscape (Fig. 12.2). Thus, in a boreal forest

Figure 12.2. Repeated pattern of an agricultural landscape in Minnesota. Farmsteads, woodlots, roads, tiny streams, and various cultivated fields form interacting clusters of ecosystems found in similar form throughout the landscape, a useful characteristic in scientifically delimiting and studying a landscape. (Photo courtesy of USDA Soil Conservation Service.)

landscape the cluster might include spruce-fir (*Picea-Abies*) woods, stream corridor, bog, rock outcrop, and aspen *(Populus)* stand. The definition also indicates that the ecosystems of the cluster are interacting. Thus, animals, plants, water, mineral nutrients, and energy are flowing from one ecosystem to another in the cluster. Each ecosystem is both a source and a sink for different moving objects. An ecosystem not only provides objects that affect neighboring systems, but in a real sense is molded and controlled by the accumulation of objects arriving from the surroundings (Hills 1974).

The ethics of isolation now begin to come into focus. In such a tightly interlocking spatial system, an action in one ecosystem cannot be isolated there. The action directly or indirectly affects each ecosystem of a cluster, and thus modifies the cluster of ecosystems as a whole.

Let us briefly consider several other principles in the developing theory of landscape ecology (Forman and Godron 1981, 1986). A later focus of this chapter is the spread of disturbances (Pickett and White 1985) in a landscape. What is a disturbance? Disturbances are events that cause a significant change in the existing pattern in a system. The event is generally an excess (sometimes a decrease) in the level of objects (such as animals, plants, water, mineral nutrients, heat) entering an ecosystem. Fire, blowing sand, or a herd of herbivores may be a disturbance. *Thus, the existence, intensity, and frequency of disturbances within an ecosystem often depend greatly on the neighboring ecosystems of a cluster.*

Continuing with the list of landscape ecology concepts, *many animals require two or more ecosystems to survive and reproduce.* Commonly an animal feeds in one, gets water in a second, and rests or raises young in a third (Fig. 12.3; Thomas 1979; Collins and Urness 1983; Wiens 1985). Here the combination of three ecosystems or habitats is critical, and the effects of a human action in one cannot be isolated in it, but are felt by animals in all three. A corollary to this principle of animals requiring two or more ecosystems, is that *the ecosystems required cannot exceed a certain distance apart to be suitable for use* by the animals. If one ecosystem is widely separated, it is energetically inefficient,

Figure 12.3. Bull elk (*Cervus canadensis*) on winter range in Yellowstone National Park, Wyoming. This species uses several different ecosystems both through the diurnal cycle and the seasonal cycle. Valleys serve as critical conduits between low-elevation winter range and high-elevation summer range for many mammals worldwide. Human activities often threaten the connectivity of such corridors. (Photo courtesy of USDA Soil Conservation Service.)

or predator susceptibility is too high, for effective use (Krebs et al. 1974; Charnov 1976).

The landscape also includes corridors that act as conduits for moving objects. For example, wind carrying dust and seeds is channeled along road corridors and the borders of hedgerows, and animals move along stream and powerline corridors. *With more conduits present in a system, it is less likely that an alteration can be isolated at one location.* In effect, a change at one spot will probably spread.

Indeed, in a system containing conduits, the conduit itself is a critical spot in terms of flows (Van der Zande et al. 1980; Forman 1983). If a gap, break, or narrows exists in the corridor, or if a disturbance occurs at a point along the corridor, that point normally forms an effective bottleneck. Objects flowing along the corridor simply cannot pass, or do so at a reduced rate (Fig. 12.3). In addition, objects crossing a corridor from one side to its opposite side are affected in a major way. Those moving by locomotion typically exert caution and slow down in passing through a gap, whereas objects transported by mass flow such as wind accelerate in passing through a gap. Nevertheless, in all cases *the corridor gap is a key control point* which responds to the effects of an action elsewhere, and transcribes or propagates the effects onward, often in different form.

Long-distance movement of migrating animals is a familiar case of an action at one spot causing a domino effect far away. Such animals use a sequence of habitats as rest stops or stepping stones as they move across the landscape (Fig. 12.2). *The loss of one or two rest stops in this chain is quite effective in drastically altering the distribution of the species, and in so doing affecting the other rest stop ecosystems.*

These concepts of a cluster of ecosystems, flows of objects and disturbances within the cluster, animals requiring two or more ecosystems, conduits spreading objects, corridor gaps acting as control valves, and long-distance movement progressively along rest stops, lead inexorably to the *ethics of isolation.* Simply stated, in land use decisions and actions it is unethical to evaluate an area in isolation from its surroundings or from its development over time. Ethics impel us to consider an area in its broadest spatial and temporal perspectives.

One may object that because everything is interrelated, it is impractical or idealistic to have to consider everything in making a decision or taking an action affecting the land. However, landscape ecology can build on the principles of geography and ecology to provide highly usable guidelines. From geography, everything is interrelated, but near objects are more related than distant objects. From ecosystem science, energy and mineral nutrients flow from one object to another within or between ecosystems. From behavioral science, because species find certain habitats more suitable or hospitable, some locomotion-driven movements tend to be directional toward patches of similar types. We may integrate these concepts into the following statement. *All ecosystems are interrelated, but flows between ecosystems decrease with distance; in addition, certain locomotion-driven flows are channeled between similar ecosystems.*

Note that flows may include typical ecological phenomena such as animals, heat energy, and mineral nutrients, as well as social phenomena such as innovations and information (Forman and Godron 1986). In short, this spatial flow principle provides a practical guideline for selecting the potential spatial interactions which must be evaluated in landscape management and land use.

An example of this primary focus on the surroundings, rather than on the characteristics of a site, comes from the astute writer and observer of nature, Henry David Thoreau. During his teaching years, 1838 to 1841:

> . . . Henry Thoreau called attention to a spot on the river-shore, where he fancied the Indians had made their fires, and perhaps had a fishing village . . . (the students) drew the boat to shore and moved up the bank a little way. 'Do you see,' said Henry, 'anything here that would be likely to attract Indians to this spot?' One boy said, 'Why, here is the river for their fishing'; another pointed to the woodland near by, which could give them game. 'Well, is there anything else?' pointing out a small rivulet that must come, he said, from a spring not far off, which could furnish water cooler than the river in summer; and a hillside above it that would keep off the north and northwest wind in winter. (Sanborn 1917).

No one can escape isolation ethics. Land owners, landscape architects, ecologists, politicians, foresters, environmentalists, wildlife biologists. . . . all make decisions and take actions which cannot be isolated in a single spot or ecosystem (Hills 1974). No area has an uncrossable boundary. It is unethical to think a boundary is impenetrable and an action can be isolated.

12.2 Heterogeneity and the Spread of Disturbance

12.2.1 Examples

The role of landscape heterogeneity in the spread of disturbance was addressed in the 1986 symposium on this topic. Yet, the number of examples and the amount of evidence to evaluate the phenomenon are limited. This is surprising, considering the immense potential significance of the subject. Nevertheless, a number of representative examples is mentioned here to illustrate the range of evidence and the opportunity for research.

1. The southern Labrador coniferous forest is composed of a matrix of spruce and fir *(Picea, Abies)*, with patches of aspen *(Populus)*, recently burned areas, and fens. Young aspen stands are essentially inflammable and thus form distinct barriers to the spread of wildfire. Fire spread also depends on the shapes and configuration of fens, some of which in turn are dependent on the previous coniferous forest patchiness and *Sphagnum* development. (David Foster, symposium presentation; Foster 1983).
2. In the Rocky Mountains pine bark beetles spread progressively through conifer forest killing mature trees. Heterogeneity in the area is caused by variations in substrate conditions, including soil moisture and geochemistry,

and by disturbance, including fire and bark beetle outbreaks. The greater the landscape heterogeneity, the less the beetle spreads, and the smaller the size of future disturbance patches in the landscape. Differences in substrate geochemistry (reflected in the trees) enhance or inhibit beetle spread. Landscape heterogeneity and pest spread are linked in a negative feedback loop: pest spread increases heterogeneity and heterogeneity decreases pest spread (Knight, Chapter 4 this volume; David A. Kovacik, symposium presentation).

3. Fire and ungulate herds are concentrated in open grassy productive areas on barrier beach landscapes of the North American Atlantic coast. Both spread as disturbances from the open areas into adjacent relatively unproductive woody areas. In this case greater landscape heterogeneity (i.e., the presence of open areas) enhances the source and spread of disturbance. (Turner and Bratton, Chapter 5 this volume).

4. Fire suppression leads to increased forest homogenization in North American conifer forests. This in turn enhances the spread of fire or bark beetles. (Knight, Chapter 4 this volume; Don D. Despain, symposium presentation; Little 1979; Forman and Boerner 1981).

5. A mathematical model suggests that if the difference between a patch and the surrounding matrix is large, species from one do not disperse far into the other. However, if the patch-matrix difference is small, species from one will invade far into the other. The model also generates the hypothesis that species characteristic of the interior of open areas will cross a greater distance of forest than forest interior species will cross open areas. (David Ludwig, symposium presentation).

6. In the Serengeti Plain of East Africa the primary causes of spatial pattern at the local plant community level are levels of sodium and termites. Local plant community pattern in turn exerts a primary control over spread of the two proximal causes of pattern at the landscape level, fire and mammal activity. (Joy Belsky, symposium presentation).

7. Some ecosystems serve as sinks, absorbing low levels of certain air pollutants. Thus, the spread of this pollution depends on landscape heterogeneity or the distribution of effective sinks. (Bormann, Chapter 3 this volume).

8. Large patches may make their own weather. This weather in turn may increase the persistence of the patches. But the meteorologic conditions also spread downwind to significantly alter neighboring ecosystems. (Meentemeyer and Box, Chapter 2 this volume; Miller 1984).

9. In the Douglas-fir (*Pseudotsuga*) forests of the Pacific Northwest disturbances such as blowdowns, fire, and certain pest damage are enhanced by, and highly sensitive to, edges between clearings and forest. Thus, the shapes of patches, reflecting the total length of edge present, are key indicators of disturbance susceptibility. Edges also act as barriers to the spread of disturbance, so no disturbance covers an entire landscape mosaic. (Jerry F. Franklin and Richard T.T. Forman, symposium presentation; Thomas A. Spies, Frederick J. Swanson, and Miles A. Hemstrom, symposium presentation).

12.2.2   Principles

Additional examples could be mentioned, but these illustrate that landscape heterogeneity may indeed play a key role in the spread of disturbance (Fig. 12.4). *In general, greater heterogeneity inhibits disturbance spread.* Nevertheless, the data base barely exists, exceptions are suggested, and opportunities for important empirical or mathematical study are rampant.

The combination of emerging landscape ecology principles and the above examples suggests at least two hypotheses or concepts that may help elucidate the role of heterogeneity in the spread of disturbance.

The spread of disturbance decreases as boundary-crossing frequency increases. This concept was discussed in terms of animal movement (Forman and Godron 1986), but also appears applicable to disturbance spread in a landscape. Implicit here is that most edges (though not all) act as barriers or filters

Figure 12.4. Fire spreading through coniferous forest in Washoa County, Nevada. Rock outcrops, clearings, roads, and other landscape elements contribute to landscape heterogeneity and decrease the probability of disturbance spread. (Photo courtesy of USDA Soil Conservation Service.)

to the spread of disturbance, due to the special structure or combination of species present in an edge. Boundary-crossing frequency may be a more simplified and direct indicator or measure of disturbance spread than landscape heterogeneity. In measuring heterogeneity one might, for example, record the number and relative abundance of types of ecosystems encountered along sample lines, and calculate an index of heterogeneity (or information). Alternatively, one could record the number of boundaries between ecosystems along the lines. Note that the information recorded in the two approaches is different. Nevertheless, the latter may be preferable for estimating disturbance spread probability, because it is more spatially explicit (being based on the particular configuration of corridors and patch shapes through which a disturbance must spread), and is easier to calculate and communicate (being a single variable rather than a synthetic index of two nonindependent variables).

Movement between patches at opposite ends of a corridor is doubtless enhanced when the corridor is wide and continuous (Van der Zande 1980; Forman 1983). Thus, *a corridor connecting patches at each end acts as a barrier or filter to movement between the patches, especially when the corridor is narrow or has breaks (i.e., low connectivity).* Furthermore, we can consider the role of a corridor relative to the patches on each side of it. *A corridor separating patches on opposite sides of it acts as a barrier or filter to movement between the patches, especially when the corridor is wide or continuous (i.e., high connectivity).* Note that in either case the corridor may be an effective bottleneck or control mechanism for the land manager to inhibit or arrest the spread of disturbance.

In short, landscape heterogeneity, as measured by the types and relative abundance of ecosystems present, can be expected to affect the spread of disturbance. However, landscape configuration, that is, the spatial juxtaposition of patches, corridors, and matrix, is doubtless a more sensitive and precise controller of disturbance spread.

Finally, I think the ideas in this section manifest a clear linkage and reinforcement of the isolation ethics message. The spread of a disturbance depends not only on a particular spot or ecosystem, but on the spatial configuration of ecosystems surrounding it in a landscape.

12.3   The Growth of Landscape Ecology

In addition to isolation ethics and the spread of disturbance, the third major topic weaving this book and the 1986 symposium together was landscape ecology and its development. Building on its early embryological phase in the 1960s, landscape ecology has undergone an impressive surge in the 1980s, particularly in Eastern and Western Europe and in North America. Several international meetings have brought scholars of diverse disciplines together and galvanized their common interest in understanding the ecology of a landscape. These include a series of symposia in Czechoslovakia (CSSR) (e.g., Proc. Vlth Int. Symp. Probs. Landscape Ecol. Research 1982; Preobrazhensky 1984), a congress in The Netherlands (Tjallingii and de Veer 1981), a workshop in the United States (Illinois) (Risser et al. 1984), and a seminar in Denmark (Brandt and Agger 1984). In 1986 the first North American symposium for landscape ecology was held in Athens, Georgia, which brought forth a wide range of data, concepts, and guidelines for this developmental phase of the field. Some are summarized and embellished in the following section.

12.3.1   Broad-Scale and Fine-Scale Approaches

Landscape ecologists were encouraged to think broadly in addressing nine challenges (Risser, Chapter 1 this volume). Think boldly; don't just knock off corners. Reject old ecological myths. Look for generality rather than specificity. Address the scaling issue. Focus on patches, their boundaries and interactions. Use new powerful analytical techniques. Determine the role of geography here. Use our ecological understanding of landscapes in land-use decisions. Build on the concepts of this symposium.

Yet, coupled with these challenges is the critical need for careful detailed research that builds a solid foundation of landscape ecology theory. A wealth of examples was presented at the symposium. Nitrogen moves between ecosystems in the tundra (G.R. Shaver, symposium presentation). Beaver *(Castor)* both create and maintain landscape heterogeneity (Remillard et al., Chapter 6 this volume). Farmers' perceptions of the land strongly determine agricultural landscape structure (Nassauer and Westmacott, Chapter 11 this volume). Experimental studies of small mammal movements demonstrate linkage patterns among woods, fencerows, and fields (G. Merriam, and J.F. Wegner, symposium presentation). A shifting mosaic in mountainous coniferous forest is usually not in steady state (D. Kovacik, symposium presentation; Knight, Chapter 4 this volume). Woodlots appear and disappear over time while an agricultural landscape remains constant (Sharpe et al., Chapter 8 this volume; Sharpe et al. 1981). Boundaries between woods and conservation tillage (no till) are sediment sinks (Odum, E.P., symposium presentation). Landscape architects change landscapes and create curves, gradients, and texture (Morrison, Chapter 9 this volume).

A combination of broadly conceived studies of large landscape tracts or whole landscapes, plus detailed studies of patterns, processes, and changes involving two or a few ecosystems, will provide the grist and foundation for rapid development of the field. If the existence of synergisms between the two approaches is recognized by all involved, the often sterile and debilitating conflict between holists and reductionists can be bypassed.

12.3.2   Pitfalls and Cairns

Scale is a frequent topic in discussions of landscapes. I find it useful to consider a region containing whole landscapes as one level of scale, with each landscape being composed of individual relatively homogeneous ecosystems at the next finer level of spatial scale (Forman and Godron 1986). This approach is useful because it meshes and integrates well the dictionary definitions of a landscape, geographers' concept of a landscape, the usual methodological approach for ecosystem study, and most important, ready communication with the public and decision makers. With this approach to the problem of scale, we can combine the central principles of landscape ecology with principles from other areas of knowledge such as micrometeorological patterns, trophic structure, biogeography, and animal behavior to understand the ecology of a landscape. Analogously, to understand how a vertebrate body works, we must draw on information at finer levels of scale including cell structure, hormone synthesis, and neural impulse transmission.

Indeed one may ask how many of the papers at the 1986 symposium were really landscape ecology, because most were designed and carried out in other contexts, such as population ecology, ecosystem ecology, and biogeography. Some were; some weren't. Landscape ecological studies explicitly include the spatial relationships of more than one ecosystem. Rather than putting on blinders or drawing imaginery boundaries around the subject, the keys to progress are the goals set, the questions asked, and the answers needed (Margalef 1968; Allen and Starr 1982). Here the focus is clearly on how a landscape mosaic is structured ecologically, how this spatial structure affects the functioning or flows, and how the mosaic changes over time. In approaching these goals, so far we have found dealing with species, mineral nutrients, and energy in relation to patches, corridors, and matrix to be a highly useful conceptual framework. Nevertheless, there is nothing sacred about the framework, and a hallmark of the landscape ecology symposia has been ecumenism, a welcoming of related disciplines.

In such a milieu it is not surprising that there has been extensive terminological debate over the years. What is a landscape? What do we call the basic units within a landscape? What is disturbance? What does scale mean? What is landscape ecology? Yet, the 1986 symposium was hearteningly free of such discussion. A recognition that a new frontier or set of questions and answers is upon us seemed infinitely more interesting than definitional debate, even though several important terms remain troublesome. Often we heard phrases like, "a new twist," or "a new way of looking at it," or "a new framework" for posing questions and research.

Landscape ecology poses different questions and provides different answers than traditional areas of ecology, and offers concrete applications for environmental issues. Nevertheless, although we may be part of the solution, at times we may also be part of the problem. Are there landscape architects who follow the client's money rather than the landscape's ecology? Or ecologists who consider their general research interest to be the only important area of ecology, making the field hopelessly fragmented? Or foresters or wildlife biologists who follow rather than create political winds? And the list could go on. Doubtless these are the minority, and the bulk of each group is committed to improving landscape quality. Indeed, it appears that landscape ecologists are in a unique position to be part of the solution. Their basic research results can be applied to land use problems facing humankind with as short a lag as in virtually any other field.

As landscape ecology moves into its growth phase, pitfalls exist as noted. Yet, cairns left by older fields are also evident to help bypass these pitfalls.

12.3.3   The Catalysis of Landscape Ecology

The state of our knowledge accelerates when instructors teach a subject and students grapple with its issues and principles. By 1987 to 1988 every university should offer a course in landscape ecology. It could be an ecology course in a biology program, develop in a number of other departments, or emerge in its own right as an interdisciplinary course drawing from several fields. Indeed, it may often begin jointly with instructors from two (or more) fields.

Research is a lifeblood of a field. Here again we can learn from subjects that began as interdisciplinary subjects, remained as the overlap area of other disciplines, and eventually withered. A body of landscape ecological research developing its own principles and theory, in consort with principles from related fields, appears to be the essential ingredient for long-term development. Research also requires funding. Good innovative research of course attracts funding. But in addition, we must demonstrate to funding sources that the promise of significant results is real (Risser et al. 1984).

Finally a user clientele for landscape ecology research must be cultivated. Research results will naturally build the body of theory, but also should be put promptly to use in landscape planning and management. Here we can readily capitalize on the strengths of the landscape approach. For example, what landscape structure or spatial configuration of ecosystems will concurrently optimize soil conservation, biological diversity, wildlife populations, visual quality, and other ecological characteristics of interest? Landscape ecology is an ideal foundation for park planning and management. Each park is a heterogeneous tract of land containing several ecosystems and plenty of people. Every large watershed, every public land, and indeed every township should have a landscape ecological plan. Decision makers and the public will soon come to realize they can communicate readily with foresters, geographers, landscape architects, wildlife biologists, ecologists, and others using the landscape ecology paradigm.

12.4   Conclusion

It is unethical to consider an area in isolation from its surroundings or from its development over time. Landscape ecology can play a key catalytic role in decreasing the gulf between government and economic actions and the demands of the land ethic. If deterioration of the landscape is to be arrested, the ethics of isolation must gnaw at our individual and social consciences.

Evaluating the role of landscape heterogeneity for the spread of disturbance has provided conceptual insights and land use applications for landscape ecology. In general, greater heterogeneity decreases the spread of disturbance. However, landscape structure or the configuration of ecosystems, which determines boundary-crossing frequency, is probably a better predictor of disturbance spread.

Landscape ecology has now emerged and is entering its growth phase. Concurrent broad-scale and fine-scale approaches offer an explicit synergism and stimulus to the field. Pitfalls evident in the development of older fields can be avoided. Teaching courses in the subject, catalyzing research to build its own body of theory, and developing a clientele for the research results will accelerate this growth phase of landscape ecology.

Acknowledgments

I warmly thank Bruce T. Milne, Jean M. Hartman, Frank B. Golley, Monica G. Turner, and Steward T.A. Pickett for comments on this manuscript, and Edward O. Wilson for discussion and insight on ethics.

References

Allen, T.F.H., Starr, T. 1982. *Hierarchy Perspectives for Ecological Complexity.* University of Chicago Press, Chicago.

Brandt, J., Agger, P. (eds.) 1984. *Proceedings of the First International Seminar on Methodology in Landscape Ecological Research and Planning,* pp. 118, 150, 153, 171, 235, 5 Vols. Roskilde Universitetsforlag GeoRuc, Roskilde, Denmark.

Charnov, E.L. 1976. Optimal foraging theory: Attack strategy of a mantid. *Am. Naturalist* 110:141–151.

Collins, W.B., Urness, P.J. 1983. Feeding behavior and habitat selection of mule deer and elk on northern Utah summer range. *J. Wildl. Mgmt.* 47:646–663.

Forman, R.T.T. 1983. Corridors in a landscape: Their ecological structure and function. *Ekologia (CSSR)* 2:375–387.

Forman, R.T.T., Boerner, R.E.J. 1981. Fire frequency and the Pine Barrens of New Jersey. *Bull. Torrey Bot. Club* 108:34–50.

Forman, R.T.T., Godron, M. 1981. Patches and structural components for a landscape ecology. *BioSci.* 31:733–740.

Forman, R.T.T., Godron, M. 1986. *Landscape Ecology.* Wiley & Sons, New York.

Foster, D.R. 1983. The history and pattern of fire in the boreal forest of southeastern Labrador. *Can. J. Bot.* 61:2459–2471.

Hills, G.A. 1974. A philosophical approach to landscape planning. *Landscape Planning* 1:339–374.

Krebs, J.R., Ryan, J.C., Charnov, E.L. 1974. Hunting by expectation or optimal foraging? A study of patch use by chickadees. *Animal Behav.* 22:953–964.

Leopold, A. 1949. *A Sand County Almanac.* Oxford University Press, New York.

Little, S. 1979. Fire and plant succession in the New Jersey Pine Barrens, pp. 279–314. *In* R.T.T. Forman (ed.), *Pine Barrens. Ecosystem and Landscape.* Academic Press, New York.

Margalef, R. 1968. *Perspectives in Ecological Theory.* University of Chicago Press, Chicago.

Miller, D.H. 1984. Ecosystem contrasts in interaction with the planetary layer. *GeoJ.* 83:211–219.

Pickett, S.T.A., White, P.S. (eds.) 1985. *The Ecology of Natural Disturbance and Patch Dynamics.* Academic Press, New York.

Preobrazhensky, V.S. 1984. International symposium on landscape ecology. *Soviet Geogr.* 6:453–463.

Proceedings of the VIth International Symposium on Problems in Landscape Ecological Research. 1982. Institute for Experimental Biology and Ecology, Bratislava, Czechoslovakia.

Risser, P.G., Karr, J.R., Forman, R.T.T. 1984. *Landscape Ecology. Directions and Approaches*. Illinois Natural History Survey, Special Publ. no. 2. Champaign, IL.

Ruse, M., Wilson, E.O. 1985. Moral philosophy as applied science. *Philosophy* 61:173–192.

Sanborn, F.B. 1917. *The Life of Henry David Thoreau Including Many Essays Hitherto Unpublished and Some Accounts of His Family and Friends*. Houghton Mifflin, Boston.

Sharpe, D.M., Stearns, F.W., Burgess, R.L., Johnson, W.C. 1981. Spatio-temporal patterns of forest ecosystems in man-dominated landscapes of the eastern United States, pp. 109–116. *In* S.P. Tjallingii, and A.A. de Veer (eds.), *Perspectives in Landscape Ecology*. Centre for Agric. Publ. and Documentation, Wageningen.

Thomas, J.W. (ed.) 1979. *Wildlife Habitats in Managed Forests: The Blue Mountains of Oregon and Washington*. U.S. Dept. Agric., Forest Service, Agriculture Handbook 553.

Tjallingii, S.P., de Veer, A.A. (eds.) 1981. *Perspectives in Landscape Ecology*. Centre for Agricultural Publication and Documentation, Wageningen.

Van der Zande, A.N., der Keurs, W.J., van der Weijden, W.J. 1980. The impact of roads on the densities of four bird species in an open field habitat—evidence of a long-distance effect. *Biol. Conserv.* 18:299–321.

Wiens, J.A. 1985. Vertebrate responses to environmental patchiness in arid and semiarid ecosystems, pp. 169–193. *In* S.T.A. Pickett, and P.S. White (eds.), *The Ecology of Natural Disturbance and Patch Dynamics*. Academic Press, New York.

Wilson, E.O. 1984. *Biophilia*. Harvard University Press, Cambridge, MA.

Index